Modeling and Optimization
in Software-Defined Networks

Synthesis Lectures on Learning, Networks, and Algorithms

Editor
Lei Ying, *University of Michigan, Ann Arbor*

Editor Emeritus
R. Srikant, *University of Illinois at Urbana-Champaign*

Founding Editor Emeritus
Jean Walrand, *University of California, Berkeley*

Synthesis Lectures on Learning, Networks, and Algorithms is an ongoing series of 75- to 150-page publications on topics on the design, analysis, and management of complex networked systems using tools from control, communications, learning, optimization, and stochastic analysis. Each lecture is a self-contained presentation of one topic by a leading expert. The topics include learning, networks, and algorithms, and cover a broad spectrum of applications to networked systems including communication networks, data-center networks, social, and transportation networks. The series is designed to:

- Provide the best available presentations of important aspects of complex networked systems.

- Help engineers and advanced students keep up with recent developments in a rapidly evolving field of science and technology.

- Facilitate the development of courses in this field.

iv

Modeling and Optimization in Software-Defined Networks

Konstantinos Poularakis, Leandros Tassiulas, and T.V. Lakshman

ISBN: 978-3-031-01254-9   paperback
ISBN: 978-3-031-02382-8   ebook
ISBN: 978-3-031-00246-5   hardcover

DOI 10.1007/978-3-031-02382-8

A Publication in the Springer series
SYNTHESIS LECTURES ON LEARNING, NETWORKS, AND ALGORITHMS

Lecture #27
Editor: Lei Ying, *University of Michigan, Ann Arbor*
Editor Emeritus: R. Srikant, *University of Illinois at Urbana-Champaign*
Founding Editor Emeritus: Jean Walrand, *University of California, Berkeley*
Series ISSN
Print 2690-4306   Electronic 2690-4314

Modeling and Optimization in Software-Defined Networks

Konstantinos Poularakis
Yale University

Leandros Tassiulas
Yale University

T.V. Lakshman
Nokia Bell Labs

SYNTHESIS LECTURES ON LEARNING, NETWORKS, AND ALGORITHMS
#27

ABSTRACT

This book provides a quick reference and insights into modeling and optimization of software-defined networks (SDNs). It covers various algorithms and approaches that have been developed for optimizations related to the control plane, the considerable research related to data plane optimization, and topics that have significant potential for research and advances to the state-of-the-art in SDN.

Over the past ten years, network programmability has transitioned from research concepts to more mainstream technology through the advent of technologies amenable to programmability such as service chaining, virtual network functions, and programmability of the data plane. However, the rapid development in SDN technologies has been the key driver behind its evolution.

The logically centralized abstraction of network states enabled by SDN facilitates programmability and use of sophisticated optimization and control algorithms for enhancing network performance, policy management, and security. Furthermore, the centralized aggregation of network telemetry facilitates use of data-driven machine learning-based methods. To fully unleash the power of this new SDN paradigm, though, various architectural design, deployment, and operations questions need to be addressed. Associated with these are various modeling, resource allocation, and optimization opportunities. The book covers these opportunities and associated challenges, which represent a "call to arms" for the SDN community to develop new modeling and optimization methods that will complement or improve on the current norms.

KEYWORDS

software defined networks, virtual network functions, network optimization, learning in networks, fault management

Contents

Preface

Network modeling and optimization are well-established topics with the fundamental goal of operating networks efficiently. Network programmability, over the past ten years, has transitioned from research concepts to more mainstream technology that is useful for making networks flexible and adaptable to changing application and provider needs. Various problems have been studied in this context, often tailored to the specific needs of the different types of networks and supporting applications. Interestingly, the range of these problems has been broadening perceptibly in recent years, which can be largely attributed to the proliferation of novel network technologies amenable to programmability, which radically changed the way modern networks operate. Among these technologies, we focus in this book on the key Software Defined Network (SDN) technology and explain how the various problems and optimization methods can be revisited with the fresh perspective and new opportunities that SDN brings.

We begin the book with a brief overview of SDN before discussing various algorithms and approaches that have been developed for optimizations related to the control plane. Following this, we provide an overview of the considerable research related to data plane optimization and cover the key ideas. Finally, we present a sampling of topics that have significant potential for research and advances to the state-of-the-art in SDN. The presented taxonomy of works and ideas can be useful to network researchers and engineers, as well as serve as a quick reference for students to become familiar with this field.

Konstantinos Poularakis, Leandros Tassiulas, and T.V. Lakshman
June 2021

Acknowledgments

We are grateful to reviewers of early drafts of this material for their constructive comments. We thank Mike Morgan of Morgan & Claypool for his encouragement and his help in getting this text reviewed and published.

The work of K. Poularakis and L. Tassiulas was supported by the U.S. Army Research Laboratory and the U.K. Ministry of Defence under Agreement Number W911NF-16-3-0001, the U.S. Army Research Office under Agreement Number W911NF-18-10-378 and the U.S. Office of Naval Research under Grant N00173-21-1-G006.

Konstantinos Poularakis, Leandros Tassiulas, and T.V. Lakshman
June 2021

CHAPTER 1

Introduction

Modern communication networks contain a large number of devices from typical routers and switches to various types of middleboxes like firewalls, network address translators (NATs), and load-balancers. Network administrators are responsible for configuring these devices to realize policies for managing the information flowing through the network. These policies need to be adaptable to network events and time-varying network conditions. Moreover, they need to support the increasingly stringent networking requirements stemming from new applications' traffic such as Augmented and Virtual Reality (AR/VR) multimedia streams [1] and delay sensitive Internet of Things (IoT) traffic [2]. Such application requirements along with massive-scaling of endpoints raise new challenges in network management and automation.

The above challenges are being addressed by increased use of high-level abstractions in network management as with intent-based networking [3]. Policies use high-level abstractions for specifying intended behavior with automated tools to translate to lower-level network device configuration thereby reducing the need for manual low-level configuration and the possibility of introducing configuration errors. This trend toward automated network management in complex hard-to-manage network infrastructure, and in the face of increasingly complex network application needs, has catalyzed the advent of new paradigms in networking.

Software Defined Network (SDN) technology is undoubtedly the most popular example of this new trend and addresses the emerging network automation needs. The aim of SDN is to make the network programmable and this way facilitate the implementation of automatic and adaptable network operations that can better meet the application needs. In SDN, control decisions are decoupled from the network devices that actually forward the traffic in the network. The network intelligence is logically centralized in software-based controllers (control plane), and the network forwarding devices (data plane) can be programmed by the controllers using an open programming interface such as OpenFlow [4]. This is a key architectural difference compared to traditional networks in which both the control and data plane functionalities are embedded into the network forwarding devices.

SDN has been successfully deployed in both Data Centers (DCs) and Wide-Area Networks (WANs) [5]. Its use also encompasses the new generation wireless (5G) and IoT networks [6]. Several factors have contributed to the SDN adoption. First, open interfaces like OpenFlow promote network programmability facilitating the rapid introduction in the network of new services and capabilities. Second, a broad range of SDN-enabled hardware devices are available as well as open-source software control plane (e.g., OpenDaylight, ONOS) and data

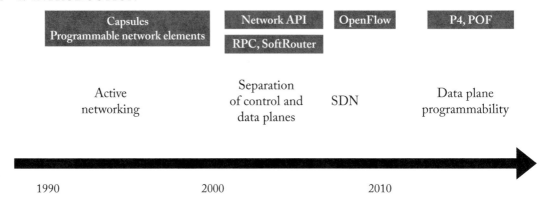

Figure 1.1: History of programmable networks. From active networking research in the mid-1990s to the separation of control and data planes in early 2000s and from there to the birth of OpenFlow/SDN concept in 2008 and the newer generation of SDN programming languages in the early 2010s.

plane (e.g., OpenVSwitch) components. Third, SDN can be deployed gradually and incrementally in parts of the network or as overlays and can co-exist with traditional network software and hardware. This reduces operational risks and makes it more realistic to deploy compared to other emerging networking architectures that require wholesale infrastructure changes.

Different aspects of the SDN paradigm have been covered by various surveys in the literature. In particular, surveys published the last few years have elaborated on the architecture elements and design challenges [7], the programming languages [8], and the security challenges [9] of SDN. However, none of these surveys has provided an extensive overview of the modeling and optimization methods to address fundamental performance and optimization challenges related to SDN. This monograph aspires to fill this gap in the literature. In the rest of this chapter, we present an intellectual history of SDN and its preceding technologies that covers over two decades, and provide a brief overview of the architecture of SDN. In the subsequent chapters, we discuss fundamental modeling and optimization issues in both the control and data plane components of the SDN network, and also highlight future research directions.

1.1   INTELLECTUAL HISTORY OF SDN

Although it may appear that SDN as a technology had a rapid incubation, this technology is in fact part of a long history of efforts in academia and industry to make networks more programmable and flexible. The history can be divided into phases that laid the foundations for many of the ideas we are seeing today, as illustrated in Figure 1.1. We briefly discuss these phases while extensive surveys are available in [10] and [11].

Starting in the mid-1990s, the *active networks* research program [12] proposed bringing programmability in the network through two major approaches: (i) capsules, an approach that placed code within packets called "in-band" to be executed at the transiting nodes in the network, and (ii) programmable network elements, an approach where the code to execute was placed directly at the nodes by "out-of-band" methods. Treating packets as programs in the capsules approach later became known as "active packets." The notion of Application Programming Interface (network API) was also introduced for exposing resources (e.g., computation and storage) on the nodes that are used as input for applying programming functions to the passing packets. Despite the considerable research activity, active networks never gathered significant attention in the industry mainly because of security and performance concerns. Nevertheless, active networking research offered some intellectual contributions and insights to network programming research and uncovered the challenges and downsides involved in executing code inside the network.

In the early 2000s, key efforts to *decouple the network data and control planes* were started and new architectures were proposed (e.g., see the SoftRouter [14] and Routing Control Platform (RCP) [13] architectures). These efforts were primarily motivated by the increasing scale of networks, and the need for flexible control based on abstracted logically centralized views of network state, enabling quicker introduction of new services and transition toward a more open architecture with use of more open interfaces. Compared to active networks, these new efforts focused on programmability in the control plane (rather than the data plane) and network-wide visibility and control (rather than device-level configuration). Accomplishing network-wide intents is facilitated by the logical centralization of control functions and abstraction of network-wide state. Shortly before SDN, Ethane defined an architecture in which a centralized controller manages policy and security in a network such as providing identity-based access control [15]. An Ethane switch consists of a flow table and a secure channel to the controller. In the more general context of SDN, the identity-based access control of Ethane would likely be implemented as one of the many applications on top of an SDN controller.

In 2008, the appearance of the *OpenFlow switch specification* [4] accelerated the trend to open interfaces and SDN. An OpenFlow switch uses a match-action paradigm. Several fields in the header of incoming packets are examined for matches to various pre-specified rules. Successful matches trigger "actions" which are packet-handling rules to be used for processing the matched packet. Different actions can be performed on the matched packets, such as forward out to a particular interface, drop, modify a header field, or move to the control plane. The advent of OpenFlow was followed by the notion of a network operating system (NOS) [18] that abstracts away lower-level network details from the logic and applications that control the network. Similar ideas of an operating system for network configuration were proposed in active networks research, yet for device-level rather than network-wide control.

While SDN/OpenFlow enabled us to "program" the data plane by populating the flow tables of the switches with rules on how to forward the traffic, they are still constrained by the pre-defined features and protocols supported by fixed function ASICs and so cannot implement, if needed, their own custom parsing functions on the packet. To address this issue, in the early 2010s, a newer generation of SDN-related solutions introduced the notion of *data plane programmability*. With this, network users can potentially write code that specifies the packet header fields and define matching and parsing semantics no longer constrained by the switch/OpenFlow specification. This is mainly fueled by data plane programming languages such as P4 [19] and POF [20]. We emphasize that while SDN/OpenFlow enabled a top-down global view of network programming, the programmable data plane concept brings the focus back to the data plane, following a rather bottom-up approach, which is in some sense similar to active networking. (This shows that "history repeats itself.")

For completeness, we note that the idea of the separation of the control from the data plane, which is usually considered a key characteristic of SDN, was already understood, and to some extent applied decades ago, as early as in the 1960s [21] and 1970s [22] in telephone networks. These networks were connection-oriented and a programmable software control separate from the data plane was introduced for controlling the opening and closing of connection points. It was actually the transition from circuit to packet switching in the mid-1990s that triggered the evolution of the Internet in the form that we know today and some years later to the adoption of network programmability and separable control and data plane concepts.

1.2   ARCHITECTURE OF SDN

Many standards organizations are engaged in SDN-related standardization, especially the Open Networking Foundation (ONF [16]) and the Internet Engineering Task Force (IETF) [17]. ONF proposed the OpenFlow architecture while the IETF proposed the ForCES (Forwarding and Control Element Separation) architecture. Both architectures follow the basic SDN principle of separation between the control and data planes and both standardize information exchange between the planes. Unlike the traditional distributed network architecture on the left of Figure 1.2 where the control of data plane devices is embedded in the devices, the SDN architecture on the right of the figure shifts the control functionality to an external software-based network entity, the SDN controller. This way, the forwarding behavior of the data plane devices is set by the controller which passes instructions to the switches through an open application programming interface (API), thereby providing a flexible programming framework for the development of new network applications.

The architecture of SDN is composed of three main layers corresponding to the data plane, control plane, and application plane, as illustrated in Figure 1.3. In addition, there are management and configuration functions associated with each layer, as we discuss in more detail below.

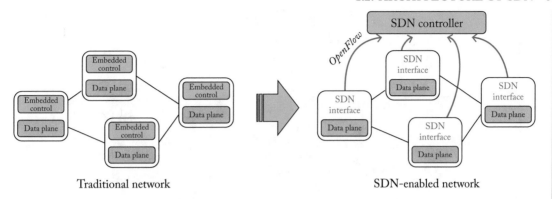

Figure 1.2: The SDN architecture decouples control logic from the forwarding hardware (data plane). The solid lines define the data plane links and the blue arrows the control plane links.

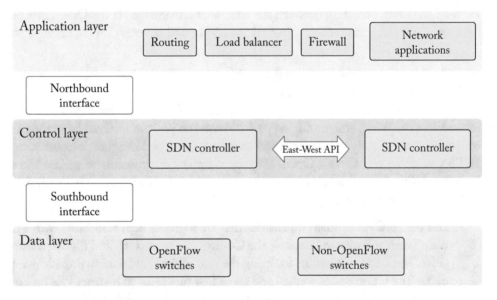

Figure 1.3: A three-layer distributed SDN architecture.

1.2.1 DATA PLANE

The data plane (DP), also known as infrastructure plane, is the lowest layer in SDN architecture. It contains the forwarding devices which can be physical switches and virtual switches. Virtual switches like Open vSwitch are software-based switches, which can run on common operating systems such as Linux. On the other hand, physical switches are implemented in hardware using either proprietary ASICs or commercial off-the-shelf forwarding engines. Virtual switches are

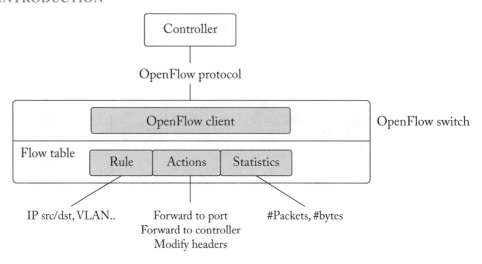

Figure 1.4: Communication between the controller and switch via the OpenFlow protocol. The flow table entries consist of matching rules, actions, and counters for collecting statistics.

more flexible to instantiate and configure. However, the physical switches have typically much higher forwarding rates than virtual switches.

The switches in the data plane, either virtual or physical, are responsible for actions such as forwarding, dropping, and modifying packets based on policies received from the controller. The controller can remotely access the switches via an open vendor-agnostic Southbound interface. IETF's ForCES and ONF's OpenFlow are examples of protocols to realize the southbound interface. While both protocols follow the same principle of separation between the control and data planes, they have some major differences.

1. IETF's ForCES defines two entities that are logically kept associated with a network element (such as a physical device): the Control Element (CE) and the Forwarding Element (FE). The FE is responsible for using the underlying hardware to provide per-packet handling. The CE executes control and signalling functions to instruct FEs on how to handle packets based on a master-slave model (CE is the master and FE is the slave). To this end, the CE uses as a building block the so-called LFB (Logical Function Block) which resides inside the FE. The CE and FE can be in close proximity to one another (e.g, same box or rack).

2. ONF's OpenFlow is recognized as the quintessential southbound protocol of SDN, and therefore we describe it in more detail in the sequel. As illustrated in Figure 1.4, an Open-Flow switch contains a flow table and an interface, the OpenFlow client, for communicating with the controller via the OpenFlow protocol. The flow table consists of multiple flow entries, each of which contains a rule that is used to check for matches with specified fields

of incoming packets headers (e.g., source or destination address). Unmatched packets are reported to the controller. A flow entry also contains a counter used to collect statistics for the particular flow such as the number of received packets or bytes and the length of the flow. Finally, a flow entry contains an action to be applied upon a match, thereby dictating how to handle the packet (forward to a port, modify a header, or forward to the controller or drop).

When an OpenFlow switch receives a packet, the packet header fields will be extracted and matched against flow entries. If a matching entry is found, the switch will process the packet locally according to the actions in the matched flow entry. Otherwise, the switch will forward an OpenFlow PacketIn message to the controller. The packet header (or the whole packet, optionally) is included in the OpenFlow PacketIn message. Then, the controller will send OpenFlow FlowMod messages to manage the switch's flow table by adding flow entries, which can be used to process subsequent packets of the flow.

We note that one of the main differences between OpenFlow and ForCES protocols is that the forwarding model of the former relies on flow tables while the latter on the LFBs. In OpenFlow the actions associated with a flow can be combined to provide greater control and flexibility for the purpose of network management. Similarly in ForCES, multiple LFBs can be combined for the same purpose. The ForCES protocol did not gain the same momentum in the industry as OpenFlow. This can be attributed to the lack of a large support community for ForCES and a number of questionable technical choices (e.g., the use of a mandatory transport protocol for interoperability among control elements).

1.2.2 CONTROL PLANE

The control plane (CP) layer is between the data and application planes. The CP is the "brain" of the network. Its main component is the logically centralized controller, which means that multiple controllers might be deployed and physically distributed across the network but their decisions should be coordinated to ensure that their collective behavior is consistent with the behavior of a single controller. The controller gathers network state information from the data plane nodes and exposes it in an abstract form to the application plane. Besides, the controller installs flow rules to the data plane devices for forwarding, blocking, modifying, or performing other actions on incoming packets. The CP does essential functionalities such as shortest path routing, traffic management, device configuration and network state management.

There are many available controllers—open-source controllers such as OpenDayLight, ONOS, NOX, POX, Floodlight, Ryu—as well as products from major equipment vendors [23]. Three communication interfaces allow the controllers to interact: southbound, northbound, and eastbound/westbound interfaces.

- The southbound interfaces (SBIs) are defined between the control plane and the data plane. Using SBIs, the data plane nodes can send network state information to the

CP and receive control policies in the form of flow rules. Examples are the popular OpenFlow (ONF) as well as several others and various proprietary SBIs.

- The northbound interfaces (NBIs) are defined between the control plane and the application plane. Using NBIs, applications can exploit the abstract network views provided by the CP to express network behaviors and requirements. This facilitates innovation, as well as automation and management of SDN networks. While standard NBIs and a common information model are being defined, different controllers in the indutry use their own NBIs.

- The eastbound/westbound interfaces are used in the case that multiple controllers are deployed. Multi-controller deployments are justified in large-scale networks where a vast amount of data flows need to be processed and use of one controller does not suffice. Multiple controllers can be also used for reliability, to avoid the single point of failure problem when only one controller is deployed. Each controller is typically responsible for only a part of the network (a subset of the switches), commonly referred to as the controller's domain. In order to provide a global network view of all domains to the upper-layer applications, the communication among multiple controllers is necessary to exchange information. The eastbound/westbound interfaces are responsible for that communication.

There exist two broad categories of eastbound/westbound protocols. The first category contains the strongly consistent protocols which strive to maintain all the controllers synchronized in all times. This is ensured by disseminating messages each time a network change (e.g., a node or link failure) happens followed by a consensus procedure. The second category contains the eventually consistent protocols which omit the consensus procedure, yet converge to a common state in a timely manner usually through periodic message dissemination. There are a variety of interfaces and no common standard. Standard routing protocols such as BGP have also been used to disseminate information between controllers.

1.2.3   APPLICATION PLANE

The highest layer in the SDN architecture is the application plane, which is composed of network applications. These applications can provide new services and perform service management and optimization. Application examples include traffic engineering, load balancer, and firewall applications, as illustrated in Figure 1.3. In general, the applications can obtain the required network state information through controllers' NBIs. Based on the received information and service requirements, the applications can implement the logic to change network policies or intents. We note that the information exposed through the NBIs is in abstract form hiding implementation details of the network infrastructure. This level of network abstraction eases network programmability, simplifies control and management tasks, and allows for innovation.

CHAPTER 2

SDN Control Plane Optimization

There exist several design choices for the centralized control plane of SDN. A key design choice is the controller placement. This involves questions like how many controllers we need, where to place in the network, and how to associate switches with controllers to form domains. Various optimization models and methods have been proposed to answer these questions targeting different metrics (e.g., latency, reliability, and cost metrics) and making different assumptions on the network conditions (e.g., static or dynamic, wired or wireless connectivity, etc.). Another problem is that the operation of controllers generates control traffic (e.g., for monitoring the switches in the data plane, and for synchronizing the network views among controllers) that can consume a significant amount of network resources. These resources could have been otherwise used to support the needs of network applications. Therefore, an important research question is how to efficiently manage the control traffic in the network taking into account the limited resource availability as well as other factors such as privacy concerns and security issues. In the following sections, we explore the problems of controller placement (Section 2.1) and control traffic management (Section 2.2) and we review the state-of-the-art models and methods and discuss the insights and lessons learned.

2.1 CONTROLLER PLACEMENT

Controller placement is one of the most thoroughly investigated problems in the SDN literature. Distributed control plane platforms like HyperFlow, ONOS, and OpenDaylight have made feasible the deployment of multiple controllers in the same network and defined mechanisms so that controllers can act as a single logically centralized entity [23]. It is therefore natural to ask how many of these controllers we need, where to place them in the network, and how to associate switches with controllers. Answering these questions will impact a wide range of network issues as we explain in the sequel.

To understand the importance of the controller placement problem we remark that due to the decoupling of the data and control planes in SDN, a switch queries its associated controller (using OpenFlow PacketIn messages) for appropriate forwarding actions (flow rule installation) every time it encounters a new flow for which the processing rules are not pre-populated in its table. Therefore, the flow setup time depends on the controller-switch latency. Consequently, the location of the controllers directly affects the flow setup time experienced by the switches which in turn affects other network issues such as the ability of the system for fast routing policy updates, quality-of-service (QoS), fault management, etc.

The pioneering work in this area was in 2012 and it showed that a single controller (if appropriately placed) suffices to meet the existing latency requirements in many networks [24]. Nevertheless, scalability and reliability concerns may still necessitate the deployment of multiple controllers, introducing new challenges and complicating the controller placement problem. Regarding scalability, we emphasize that the controller is a software process running on a machine and limited by its resources (processor, memory, access bandwidth) and therefore can manage a limited number of switches. Though running a controller on a powerful cluster will provide better scalability, it will also increase the cost of deployment and operation. Regarding reliability, the controllers as well as the connections between the controllers and switches might be subject to failures, e.g., hardware or software malfunctions. When a controller becomes unavailable, a switch will have to associate with another controller to remain operational and ensure reliability of the system.

It is clear based on the above discussion that there are many avenues for exploration of the controller placement problem. Researchers working in this area have proposed a number of metrics to measure performance. These metrics are used as objectives in optimization problems where the number of controllers and their placement depends on the specific objective one wants to optimize. As illustrated in Figure 2.1, the objectives can be classified into three broad categories; latency, reliability, and cost. Each category is further divided into multiple subcategories since there are many variants of these objectives in the literature as we explain in the sequel.

It is worth noting that the objectives of the controller placement problem are often conflicting, involving trade-offs between them. For instance, to minimize controller-switch latency it is more appropriate to place many controllers spread across the network so that each switch can find at least one controller in close distance. Yet, the deployment cost will be high in this case. On the other hand, a more compact placement where controllers are placed close to each other may be more appropriate to handle controller failures as the switches can quickly find a controller nearby to the failed one to re-assign without increasing the latency of failover much. Such problems are typically addressed by adopting a multi-objective optimization framework or by moving some objectives to the constraint list of the problem.

We emphasize that even for the same objective, the problem formulations proposed in literature are often different from each other depending on the exact modeling assumptions each work makes. These assumptions are often motivated by the application domain. For instance, in Data Center (DC) networks the traffic load of the switches can be higher than in WANs and thus more likely to overload the capacities of the controllers. Therefore, it becomes critical to add in the problem formulation constraints that limit the computation load of each controller, whereas such capacity constraints are often neglected in WANs (e.g., see [24]). The complexity of the problem and the adopted solution method highly depend on the choice of such modeling

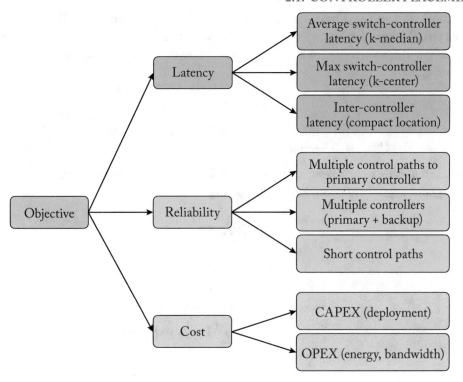

Figure 2.1: Classification of controller placement objectives.

assumptions, while clearly there is a fundamental tradeoff between model tractability and model realism.

In the sequel, we present a basic model of the controller placement problem that involves several of the metrics discussed above (Section 2.1.1). Following that, we review the state-of-the-art work on controller placement classified according to their objective: latency, reliability, and cost (Sections 2.1.2, 2.1.3, and 2.1.4). We conclude this section by presenting some more advanced methods for dynamic controller placement (Section 2.1.5) as well as emerging applications in wireless networks (Section 2.1.6).

2.1.1   BASIC MODEL

State-of-the-art work models the controller placement problem in different ways. Below, we present one of the most basic models. The network is modeled by a graph $G = (V, E)$ where V is the set of SDN switches and E is the set of network links interconnecting the switches. The notation d_{uv} denotes the distance between switches u and v which can represent the latency of the shortest path among the two switches. The notation L denotes the computation capacity of a controller, whereas λ_u denotes the control load generated by switch u which depends on the volume of flow setup requests and the type of network applications. The notation D denotes the switch-controller latency threshold. The output of the problem is the number of controllers, location, and switch-controller associations. These decision variables can be represented as follows:

$$x_u = \begin{cases} 1 & \text{if a controller is placed at switch } u \\ 0 & \text{otherwise.} \end{cases} \tag{2.1}$$

$$y_{uv} = \begin{cases} 1 & \text{if switch } u \text{ is associated with the controller at switch } v \\ 0 & \text{otherwise.} \end{cases} \tag{2.2}$$

Since the above variables are integers, we can formulate the controller placement using an Integer Programming formulation, like the following:

$$\min_{\mathbf{x},\mathbf{y}} \sum_{u \in V} x_u \tag{2.3}$$

$$\text{subject to: } \sum_{v \in V} y_{uv} = 1, \ \forall u \in V \tag{2.4}$$

$$y_{uv} \leq x_v, \ \forall u, v \in V \tag{2.5}$$

$$\sum_{u \in V} \lambda_u y_{uv} \leq L, \ \forall v \in V \tag{2.6}$$

$$d_{uv} y_{uv} \leq D, \ \forall u, v \in V \tag{2.7}$$

$$x_u, y_{uv} \in \{0, 1\}, \ \forall u, v \in V, \tag{2.8}$$

where $\mathbf{x}$ and $\mathbf{y}$ are the respective variable vectors we aim to optimize. The objective in Equation (2.3) is to minimize the number of controllers we place. This is a simple yet important objective for many reasons. First, fewer controllers require lower cost of deployment. Second, fewer controllers mean lower complexity of management and less traffic overheads for signalling and state synchronization among the controllers. The constraint in Equation (2.4) ensures that each switch u is associated with exactly one controller. Equation (2.5) guarantees that the switch u is associated with a controller v only if the controller exists (is placed). Equation (2.6) ensures that the sum of control loads of switches associated with a controller does not exceed the controller's computation capacity. Equation (2.7) ensures that each switch is associated with a controller with latency no more than the latency threshold, this way ensuring some minimum QoS level. Equation (2.8) indicates the integer nature of the optimization variables.

The above model, albeit basic, is a good example of how we can capture multiple metrics in the problem formulation. On the one hand, the objective concerned with the cost of deployment by reducing the number of controllers. On the other hand, the constraint in Equation (2.7) is concerned with the latency of controller-switch communication. These two metrics are conflicting since reducing latency requires using more controllers. The solution to the problem will find the right balance between the two metrics. Despite its simple formulation, the problem is challenging to solve. The optimization variables in vectors $\mathbf{x}, \mathbf{y}$ are integers and therefore the problem is non-convex. It can be shown that the problem is NP-Hard by reduction from the well-known vertex cover problem which is NP-Hard. Therefore, it is unlikely to find the optimal solution in polynomial time.

2.1.2   LATENCY-CENTRIC METHODS

As discussed earlier, the latency between controllers and switches is a critical metric and affects several network issues. The latency has three components: (i) the propagation latency based on the network path length between the controller and the switch; (ii) the transmission latency based on the respective port rates; and (iii) the queuing latency based on the computation capacity and current load of the controller. Each component may be more or less important depending on the application scenario of interest. For example, the propagation latency is typically the dominant factor for WANs with high link capacities that span large geographical regions across the globe, but it may be of the same order or even lower than the transmission latency for other types of networks deployed in a small region such as Data Centers (DCs). The queuing latency, on the other hand, becomes increasingly important in low-end, low-capacity networks where new flow-setup demands may overload controller capacity such as in some Internet of Things (IoT) networks.

Table 2.1 lists the state-of-the-art work addressing the controller placement problem with latency as the main objective. We classify the work according to the modelling assumptions they make—whether they optimize or assume a given number of controllers, whether they consider or neglect the capacities of the controllers, and whether they explicitly optimize the switch association decisions (y_{uv} variables in Equation (2.2)) together with the controller placement decisions (x_u variables in Equation (2.1)).

The pioneering work in [24] introduced the controller placement problem in WANs where propagation latency between controllers and switches is the most important metric. It addressed the problem as a facility location which is a long-studied problem that appears in many contexts, such as locating warehouses near factories and optimizing the locations of content distribution nodes and proxy servers in the Internet. Two variants of the latency metric were considered as objectives: (i) minimizing the average propagation latency called the k-median or k-mean problem and (ii) minimizing the worst case propagation latency called the k-center problem. Here, an assumption is that the controllers have sufficient capacity to serve the load of the switches. Therefore, from a latency perspective, it would be inefficient for a switch to associate with any controller other than the one closest to it. Therefore, we do not need to explicitly define variables to optimize the switch assignment decisions (variables y_{uv} in Equation (2.2)); they can rather be directly inferred from the controller placement decisions. Specifically, the average propagation

Table 2.1: **State-of-the-art work on latency-centric controller placement**

Ref.	Objective	Controller Number Opt.	Controller Capacity	Controller-Switch Assoc.	Method
[24]	min average propagation latency and min worst-case propagation latency	✗	✗	✗	Brute force exhaustive search to provide optimal solution
[27]	min average propagation latency and min worst-case propagation latency	✓	✗	✗	Modified affinity propagation clustering
[28]	min worst-case propagation latency	✗	✗	✗	Modified k-mean clustering
[29]	min worst-case propagation, transmission, switch processing, and queuing latencies	✓	✗	✗	Clustering-based network partition
[31]	max payoff of switches representing latency	✗	✓	✗	Clustering method utilizing cooperative game theory
[32]	min worst-case propagation latency	✗	✓	✓	Integer programming
[33, 34]	min average propagation latency	✗	✓	✓	Particle swarm optimization
[35]	min controller-switch and inter-controller latency	✗	✗	✗	Pareto-based ILP and evolutionary heuristic
[36]	min controller-switch and inter-controller latency	✓	✗	✓	Nash-bargaining game theory

latency objective is given as a function of the controller placement x_u variables only:

$$D_{avg}(\mathbf{x}) = \frac{1}{|V|} \sum_{u \in V} \min_{v \in V : x_v = 1} d_{uv} \tag{2.9}$$

and the worst-case propagation latency objective as:

$$D_{wc}(\mathbf{x}) = \max_{u \in V} \min_{v \in V : x_v = 1} d_{uv}. \tag{2.10}$$

The authors showed that we cannot minimize both the above latencies simultaneously. For a number of real networks, they measured the optimal average and worst-case latencies by trying out all possible combinations of the controllers (brute force approach). They showed that latencies reported for optimal placement of the controllers are significantly better than random placements. However, such a brute force method is computationally exhaustive and can only be used in small or medium size networks (e.g., in the order of tens of nodes). Besides, this work assumes that the number of controllers required is known beforehand.

Subsequent work addressed controller placement as a clustering problem. Clustering is the task of grouping a set of objects in such a way that objects in the same group (called a cluster) are more similar (based on some similarity metric) to each other than to those in other groups. In SDN, the objects are the switches, and a cluster is a subset of switches associated with the same controller (also called a domain). Therefore, clustering can be directly mapped to controller placement. The work in [27] used the so-called affinity propagation clustering algorithm to form clusters and estimate the number of controllers. This algorithm in its original form uses as similarity metric the negative squared Euclidean distance between objects (switches). Since the switches are generally not directly reachable in a real network topology, the authors proposed a modified version of the affinity propagation clustering algorithm in which the similiarity metric is based on the reachable shortest path instead of Euclidean distance. This metric favors the selection of clusters (and therefore also switch assignments) with small average and worst-case propagation latencies between controllers and switches. The work in [28] used a modified k-mean clustering technique again using the shortest path rather than the Euclidean distance to measure the similarity between switches. A drawback of this technique is that it requires knowledge of the number of clusters (controllers) and the initialization of the cluster centers (controller locations) which makes the algorithm sensitive to the initialization parameters. The authors compared their work with the standard k-means technique and showed an improvement by a factor of greater than 2. The work in [29] is another clustering based approach to determine the controller placement. A novel aspect of this work is a more detailed latency model that takes into account all possible components: propagation, transmission, and queuing latencies. An interesting point is the consideration of the queuing latency not only at the controller but also at the switch side, which is referred to as switch processing latency and is treated as a fourth component in addition to the other three above. When a switch sends a flow setup request to a controller, the overall latency will be the sum of the time it takes for the request message to reach the controller ("end-to-end latency") plus the time the controller takes to process the request ("queuing latency in controller"). Note how the end-to-end latency increases with every link (hop) between the switch and the controller. Specifically, let links(u, v) be the links between switch u and controller v. Then, the end-to-end latency is given by:

$$D_{end-to-end}(u, v) = \sum_{l \in \text{links}(u,v)} \left(\frac{P_l}{B_l} + \frac{d_l}{S} + DSP_l \right), \tag{2.11}$$

where P_l is the number of bits of a packet in link l and B_l is the bandwidth of the link, so P_l/B_l is the transmission latency. d_l is the distance of link l and S is the signal speed at which data travels through the medium, so d_l/S is the propagation latency. DSP_l is the processing latency of the switch at the transmitting endpoint of link l which is affected by the load and the type of the switch. Since modern SDN switches have high line rates (100 Gbps or more), the switch processing latency can be ignored. The queuing latency at the controller is more complex. The authors model it as an M/M/m queuing system, where m is the number of servers (controllers) that collaborate to serve the requests of the switches inside a part of the network under their administration. Here, the traffic intensity is termed as $\rho = \lambda/m\mu$ where $\lambda = \sum_i \lambda_i$ is the aggregate load of the switches in a part of the network and μ is the capacity of a controller. Then, we can define the steady-state limiting probabilities π_k that the system contains k customers (pending requests) given as:

$$\pi_k = \pi_0 \left(\frac{\lambda}{\mu}\right)^k \frac{1}{m!m^{k-m}}, \ k < m \tag{2.12}$$

and given that the sum of probabilities is equal to 1, we can show that:

$$\pi_0 = \left[\sum_{k=0}^{m-1} \frac{(m\rho)^k}{k!} + \frac{(m\rho)^m}{m!(1-\rho)}\right]^{-1}. \tag{2.13}$$

Based on the above, the queuing latency can be defined as:

$$D_{queue} = \frac{(m\rho)^m \rho\pi_0}{m!(1-\rho)^2\lambda}. \tag{2.14}$$

The authors proposed a clustering algorithm that allocates cluster centers based on shortest path distance between switches (computed using the Dijkstra's algorithm). Although the authors include the different latencies in their problem formulation, in the clustering algorithm, they make the simplifying assumption that the shortest path distance incorporates the end-to-end latency.

A different clustering method based on cooperative game theory was proposed in [31]. The partitioning of the network into subnetworks is modeled as a cooperative game with the set of all switches as the players of the game. The switches try to form coalitions with other switches to maximize their payoff. Here, the payoff is inversely proportional to the latency between switches favoring the selection of switches that are close one another in the same coalition. Variants of this method that try to produce partitions that are balanced in size (in case that controller capacities are at risk of being overloaded) are also proposed. The next step is to compute the Shapley value for each player (switch), which is a well-known concept in cooperative game theory. The switches with the highest Shapley values will be chosen to be the initial controller locations and the k-means clustering algorithms will be applied to them to compute the final controller locations.

It is shown that with the help of cooperative game theory better performance can be achieved compared to the pure k-means clustering algorithm.

Yao et al. [32] argued that the controller can be overloaded with many requests from the switches which can cause significant processing latency increase and packet loss. To prevent this from happening, we need to relax the policy of associating each switch with each nearest controller and, instead, opt for a policy that spreads the switches across the controllers to balance their loads. This requires explicitly defining variables for the switch assignment (variables in Equation (2.2)) and imposing a capacity constraint on them (constraint in Equation (2.6)). Following this approach, the authors extended the formulation in [24] of minimizing the worst-case latency. They defined this problem as a variant of the well-known capacitated k-center problem. They proposed an algorithm that uses integer programming to find the minimal number of controllers within a specified radius, where the radius denotes the worst-case latency, which is found using a binary search procedure. The authors claim that their algorithm, although computationally heavy, can scale for networks with tens to thousands of switches.

Gao et al. [33] proposed minimizing the average controller to switch latency subject to controller capacity constraints using Particle Swarm Optimization (PSO) algorithm. At a high level, PSO algorithm is initialized with a population of random solutions and searches for optima by updating the solution over time. In PSO, the potential solutions, called particles, "fly" through the problem space by following the current optimum particles. In this problem, each particle represents a placement instance of a given number of controllers. This approach was shown to perform better than a simple greedy placement method. Liu et al. [34] proposed another variant of PSO algorithm where elements of the particle also include the number of switches associated with each of the controllers (in addition to the placement of controllers).

Zhang et al. [35] experimentally showed that the interaction and signaling between the controllers for state synchronization can impact the reaction time perceived by the switches. The exact impact depends on the consistency protocol (e.g., strongly consistent vs. eventually consistent) used and may differ from one controller implementation to another (e.g., OpenDayLight vs. ONOS). This suggests that both the controller-switch latency and the inter-controller latency are important to ensure low reaction time perceived by the switches (while all the previous works mentioned above considered only the controller-switch latency). Motivated by this observation, an evolutionary algorithm that finds a Pareto optimal solution with respect to the average controller-switch and inter-controller latency metrics was proposed.

Finally, the work in [36] presented a formulation of the problem of minimizing the controller-switch and inter-controller latencies. While the formulation does not include explicit constraints on the capacities of controllers it does include a load balancing constraint of the following form:

$$\left| \sum_{u \in V} (y_{uv} \lambda_u) - \sum_{u \in V} (y_{uv'} \lambda_u) \right| \leq L_{diff}, \qquad (2.15)$$

where for each pair of controllers v and v', the absolute difference of the two loads cannot be larger than a threshold L_{diff}. To analyze the tradeoff between the two metrics, this work used the Nash bargaining game. Specifically, the two metrics were considered as two players. To use Nash bargaining, the so-called threat point needs to be found first which in their case is the worst-case controller-switch and inter-controller latencies. Then, the trade-off between the two latencies is modeled as an optimization problem and the threat point is computed using linear optimization.

2.1.3   RELIABILITY-CENTRIC METHODS

Disruptions in the network such as failures of controllers, switches or links can cause communication interruptions between the control and data planes in SDN. Controllers from the switches under their administration will be unable to respond to flow setup requests until communication is restored. On top of that, controllers disconnected one from the other are unable to function as a cluster, synchronize their states to be consistent and coordinate their forwarding decisions. These pathological situations lead to packet loss and performance degradation. SDN reliability considerations lead to the interest in use of reliability-centric methods.

The controller placement problem should be revisited to account for reliability. Since the management and control messages in SDN (PacketIn, FlowMod messages) are transmitted through control paths (i.e., paths connecting controllers with switches as well as controllers with each other), the reliability of the control paths directly affects SDN reliability. There are in general three approaches to improve the reliability of the control paths, as illustrate in Figure 2.1. The first approach is to provision the control paths which consist of fewer network components (switches and links) with the expectation that shorter paths will be more reliable. The second approach is to add redundancy by provisioning multiple *disjoint* control paths between the same pair of a switch and a controller. This way, and with a fast failover mechanism, communication to the controller can be quickly restored as long as at least one available control path exists. The third approach is to not only provision multiple paths but also direct these paths to multiple controllers (one primary and multiple *backup* controllers) to account also for failure of the controllers.

The controller placement should be redefined with the above approaches in mind to improve the overall reliability. Which control paths are available to provision depends on the number and location of the controllers. For example, placing many controllers widely spread across the network can help in provisioning short (with few hops) control paths since there will be always a candidate controller to pick close to every switch. Yet, the inter-controller paths will be longer (and thus possibly more vulnerable) compared to a more compact location of controllers. In addition, depending on the location of controllers in the network, it may be easier to find more disjoint paths connecting a switch to the same or different controllers.

Motivated by the above, a significant amount of work has been devoted towards designing methods for reliable controller placement. These followed the pioneering work in [24] and their aim is to extend rather than replace it. Therefore, latency is still considered to be an important metric for controller placement, and reliability is treated as an "add-on" to the problem formulation. Since the two metrics, reliability and latency, can be conflicting, most formulations use one of them as a constraint while the other as the objective.

Table 2.2 lists the state-of-the-art work that addresses the reliable controller placement problem. As in the previous table, we classify the works according to the modeling assumptions they make, i.e., whether they optimize or assume that the number of controllers is given, whether they model capacities for controllers (or assume they are infinite) and whether they optimize together with controller placement decisions the switch associations. We additionally include two columns that are specific to the reliability problem. The first column classifies works based on whether along with the controller placement (and possibly switch association) decisions they also optimize the *provisioning of control paths* or not. This is important since the shortest paths between the controllers and switches that is the default option adopted in several papers may not always be the most reliable ones. The second column classifies works based on whether they assume that the probabilities of failure of network components are known or not. Clearly, knowing the probability that each switch/link might fail can lead to more effective decisions but such information is not always available or accurate.

Known Failure Probabilities
The works in [37, 38] are among the first to consider the reliability of the controller placement. They introduced a reliability metric referred by *control path loss* δ given by:

$$\delta = \frac{1}{m} \sum_{i \in V \cup E} b_i \, p_i, \tag{2.16}$$

where $m = n + k(k-1)/2$ is the number of control paths; a single control path for each switch and its associated controller (n switches in total) and a single control path for each pair of controllers (k placed controllers, therefore $k(k-1)$ pairs of them). In the above expression, the sum is taken over all network components, switches (set V), and edges (set E). The notation p_i is the probability of failure of component i which is assumed to be known. The notation b_i is the number of control paths that are broken because of the failure of component l. The assumption in these works is that the control paths are the shortest paths (with respect to the number of hops) connecting the controllers to switches and the controllers to one another. Therefore, b_i depends on the selection of shortest paths which in turn depend on the locations of the controllers. However, this line of work did not provide any mathematical formulation with optimization variables for the controller placement decisions. Instead, they directly proposed three simple heuristic methods: (i) a random method that places the k controllers at k random switches in the network; (ii) a greedy method that ranks switches increasingly based on their failure probabilities and then iteratively places a controller to the location with the lowest path loss value in conjunction with the controllers already placed in previous iterations; and (iii) a Simulated Annealing method which is a generic probabilistic meta-algorithm. They also proposed a fourth

Table 2.2: State-of-the-art work on reliable controller placement

Ref.	Objective	Controller Number Opt.	Controller Capacity	Controller-Switch Assoc.	Control Path Opt.	Known Failure Prob.	Method
[37, 38]	min control path loss	✗	✗	✗	✗	✓	Random, Greedy, Simulated annealing heuristics, and brute force
[39]	max resilience based on cascading failure analysis	✗	✗	✓	✗	✓	Greedy heuristic
[40]	min reliability factor	✗	✗	✓	✗	✓	Clustering, Greedy heuristics
[41, 42]	min deployment and connection cost s.t. connection probability	✓	✗	✓	✗	✓	Connectivity-based heuristic
[43]	min deployment and connection cost s.t. coverage probability	✓	✗	✗	✗	✓	ILP
[44]	min # controllers s.t. latency and reliability	✓	✗	✗	✗	✗	Min-cover heuristic
[45]	max controller-switch path diversity	✗	✓	✓	✗	✗	ILP for primary CP, proximity and capacity-based heuristics for backup CP
[46, 47]	min latency, controller-less switches and load imbalance after failures	✗	✗	✗	✗	✗	Pareto-based ILP and Simulated Annealing heuristic
[48, 49]	min latency between switch and kth reference controller	✗	✓	✓	✗	✗	ILP and Simulated Annealing heuristic
[50]	min # controllers s.t. latency and controller multi-connectivity	✓	✓	✓	✗	✗	Clique-based heuristic
[51]	min average controller switch path length s.t. two paths per switch	✗	✗	✓	✓	✗	ILP
[52]	min # of controllers s.t. robust tree	✓	✗	✓	✓	✗	k-critical heuristic
[53]	min inter-controller latency both pre- and postfailure (Steiner tree)	✗	✓	✓	✓	✗	ILP

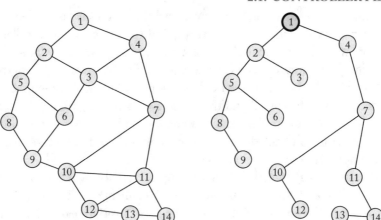

Figure 2.2: Network topology (left) and overlay tree network for communication between the controller at switch 1 and the rest switches (right) [39].

method that is rather straightforward and brute force: to generate all the possible combinations of k controllers in every possible location and return the combination with the lowest path loss value. This is clearly the optimal solution but requires non-polynomial time since there exist an exponential number of combinations to check. A limitation of these works is that the expression in Equation (2.16) double counts the broken control paths when multiple failures occur in the network since two failed edges i and i' may cause the same control path to break. This path will be double counted in both b_i and $b_{i'}$. Therefore, this model is accurate only when assuming one rather than multiple failures. On the positive side, an interesting result in these works is the exploration of the tradeoff between latency and reliability metrics. Evaluations showed that optimizing for average and worst-case latencies decreases reliability by 13.7% and 13.8%, respectively; on the other hand, optimizing for reliability increases the average and worst-case latencies by 17.3% and 13.4%, respectively.

The work in [39] introduced another reliability metric based on a *cascading failure* analysis on an interdependence graph of the network. The motivation of this work is that the switches assigned to the same controller can be connected to it using an overlay tree structure so as to save communication costs. For example, consider the overlay tree on the right side of Figure 2.2 where a single controller is placed at switch 1, while the original network topology is shown on the left side. Then, the critical point is that if a switch or edge in the overlay tree fails then the failure propagates to the descendant switches and associated edges that depend on that failed switch or edge to reach the controller. For example, in the figure, if switch 7 fails, then switches 10–14 and their associated edges will also be affected. Note that there might be an alternative network path for switches 10–14 to reach the controller through switch 9. However, detecting and establishing such a routing path is out of scope of this work. With the above reliability metric in mind, a greedy algorithm is proposed that for a given number of controllers k, first partitions the network into k sub-networks, and then selects as controller the switch in each sub-network with the maximal closeness centrality to all other switches in the sub-network. Maximal closeness centrality refers to the average of the shortest path length from the switch to every other switch in the sub-network. The rationale of using this metric is that the overlay tree will be shorter (more balanced) and therefore fewer switches will be affected by a node or edge failure. Although the controller locations and switch assignments are optimized by the above algorithm, there is no detailed discussion on how to select the most appropriate overlay tree structure to optimize the control paths.

The work in [40] argued that we can improve reliability if we provision many rather than one path between each switch-controller pair. A challenge here is how to define the reliability of a switch-controller pair as a function of the set of provisioned paths. Indeed, the number of possible paths is in general exponential to the network size, therefore even enumerating these paths can be computationally intractable. To overcome this challenge, [40] introduced a new reliability metric, called *reliability factor* that has a simple (of polynomial size) expression. This metric is intuitive and, for a given controller-switch pair, it has higher value when (i) there are more paths provisioned for this pair, (ii) the paths are shorter on average, (iii) the paths are less correlated (overlapping), and (iv) the difference degree between the lengths of the paths is higher. With this metric in mind, a k-means clustering algorithm is proposed to partition the network into sub-networks and place one controller per sub-network. To reduce computation time, a second greedy-based algorithm is proposed which sorts the switch pairs in descending order of their reliability factor value and then place controllers one-by-one in a locally optimal manner.

The work in [41, 42] made another attempt to improve reliability. Unlike [40] that provisioned multiple control paths for connecting a switch to the *same* controller, these papers suggested associating a switch to many *different* controllers. This is important since sometimes the controller itself might fail, so more controllers are needed to handle the failure. An optimization model was proposed where along with the placement of controllers the switch-controller association is decided. A main challenge here is how to define the reliability of a switch u as a function of its set of associated controllers C_u. Indeed, finding the exact reliability expression can be computationally intractable due to the exponential number of subsets of controllers and possible paths to them. To overcome this challenge, an important indicator of reliability is the number of *disjoint* paths connecting a switch u to *any* of the controllers in C_u. Let A_{u,C_u} be a set of such disjoint paths which can be found in polynomial time by solving a max flow problem (see [41, 42] for technical details). Then, we can define as reliability metric the following lower bound on the probability of having at least one operational (not broken) path between switch u and C_u:

$$R(u, C_u) = 1 - \prod_{\forall \alpha \in A_{u,C_u}} \left(1 - \prod_{\forall i \in \alpha} (1 - p_i)\right). \tag{2.17}$$

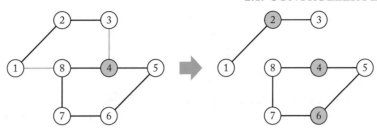

Figure 2.3: Illustrative example of controlled proportion (Cpr) reliability metric for $k = 2$ failures in [43].

In the above expression, p_i again denotes the probability of failure of network component i. The inner product will be 1 if none of the components in the path α have failed which means that the path α is still operational. Otherwise, the inner product will be 0. This expression assumes independence of failures. The external product will be 0 if at least one of the paths is operational; otherwise 0. Therefore, the overall expression will be 1 if switch u is still connected to at least one of the controllers in C_u. Having defined the above reliability metric, [41, 42] formulated the problem of minimizing the controller placement and switch controller association connection costs (which are linear functions of the respective decision variables) subject to a threshold reliability constraint $R(u, C_u) \geq R_{th}$ per switch j, where R_{th} is a constant. Then, they proposed an algorithm that heuristically ranks all switches that are candidates for controller placement according to their expected contribution to the above reliability metric. The rationale here is to prefer switches that could be useful to increase the reliability for many other switches. While this model is more general than the previous ones, the actual control paths are not optimized but only a number of disjoint paths is counted in order to compute the lower bound values $R(u, C_u)$. Besides, the proposed algorithm is again a heuristic lacking any performance guarantees.

The work in [43] introduced a new reliability metric referred as *controlled proportion* (Cpr). It represents the fraction of switches that can access at least one controller after any k edges fail. The idea is that if we allow multiple (k) failures to happen it is very likely that the network will be fragmented into multiple parts and therefore it will be impossible for some of the switches to reach any of the controllers. Consider for example the network on the left side of Figure 2.3 where a single controller is placed at switch 4. If edges $(1, 8)$ and $(3, 4)$ fail then the network is fragmented into two parts where switches 1, 2, and 3 are completely disconnected from the controller at switch 4. It can be shown that the Cpr value is $5/8$ under any $k = 2$ failures for this controller placement. In order to achieve Cpr $= 6/8$ or higher we need to place more controllers as illustrated on the right side of the figure where controllers at switches 2 and 6 are added. The problem of minimizing controller placement and controller-switch connection costs subject to a lower bound constraint on the Cpr value was formulated as a linear integer problem. The limitation of this formulation is that it enumerates all the possible scenarios of k failures which are exponential with respect to the number of edges. Specifically, the objective of the problem is given by:

$$\min \sum_{u \in V} \beta_u x_u + \sum_{s_k \in S_k} q^{s_k} \sum_{u,v \in V} \gamma_{uv}^{s_k} y_{uv}^{s_k}, \tag{2.18}$$

where S_k represents all the possible scenarios of k edge failures and q_{s_k} denotes the respective probability of a scenario $s_k \in S_k$. The notation β_u refers to the cost of placing a controller at switch u while $\gamma_{uv}^{s_k}$ refers to the cost of associating switch u to controller v which depends on the shortest path between u and v in that failure scenario s_k. Finally, x_u and $y_{uv}^{s_k}$ are the controller placement and switch assignment binary decision variables. In order to overcome the exponential representation of the objective function challenge, the authors in [43] utilize the corresponding dual programming formulation which can be solved in polynomial (cubic) time with respect to the size of the network. Although practical, this method (similar to all the above) requires knowledge of the link failure probabilities in each scenario.

Unknown Failure Probabilities

While the above works provide valuable insights into the reliable controller placement problem, they all make a rather strong assumption that *the probabilities of failures of network components (switches and/or edges) are known*. This may be a realistic assumption for some types of networks where the reliability of the hardware is well established and the failures can be predicted accurately given a certain expected workload in the network and maintenance practices by the operator. Yet, in many cases, such predictions are not possible or are inaccurate leading to inappropriate controller placement decisions. Many researchers recognized this challenge and proposed models and methods that do not require such failure probability knowledge. We review such work in the sequel.

The work in [44] defined the problem of placing the minimum number of controllers to "cover" all switches. Here, by cover we mean that a latency and a reliability constraint are satisfied. These constraints are defined as follows. Each switch is assigned to its closest controller using the respective shortest path as a control path. The delay of this path is simply the aggregate delay of all hops on this path and needs to be below an upper bound constraint. The reliability constraint is more difficult to describe. For a given controller placement and control paths between switches and controllers based on the shortest paths, the network is partitioned into sub-networks or domains, one per controller. The reliability constraint requires that in each domain the average number of control paths broken due to a single link failure will not be more than a threshold value. This approach is similar to the one in [37, 38] we described above in the sense that they all aim to limit the number of broken control paths due to a single link failure. The difference, however, is that in this work the reliability metric is in the constraints rather than in the objective and there is no assumption on the probabilities of the link failures. Since the number of possible covers is exponential to the number of nodes, a heuristic algorithm is proposed that uses ideas from the field of Particle Swarm Optimization to make the algorithm converge faster.

The work in [45] argued that we can improve reliability if we provision many rather than one path between each switch-controller pair. This is the same argument used by the work in [40] we discussed above, however this time there are no assumptions made on the probabilities of failure of the paths. The problem is modeled mathematically as an Integer Linear Program (ILP) with the goal:

$$\max \sum_{u \in V \setminus C} \sum_{c \in C} w_{uc}, \tag{2.19}$$

where variable w_{uc} counts the number of disjoint paths between switch u and controller c (in set C). The value of w_{uc} depends on the controller placement and switch assignment decisions which is modeled by adding the following two linear constraints:

$$w_{uc} \leq \sum_{v \in V} (x_{cv}) DP_{uv}, \ \forall c \in C, u \in V \setminus C \tag{2.20}$$

$$w_{uc} \leq y_{uc} \cdot \kappa, \ \forall c \in C, \ u \in V \setminus C, \ \kappa \to \infty. \tag{2.21}$$

In the above equations, following the model in [45] we slightly changed the definition of the variables x_{cv} and y_{uc} compared to Equations (2.1) and (2.2). The new variable x_{cv} indicates whether controller c is placed at switch v or not. Similarly, the variable $y_{uc} \in \{0, 1\}$ indicates whether switch u is associated with controller c or not. The first constraint upper bounds w_{uc} by the actual number of disjoint paths between switch u and the switch v that hosts the controller c. This upper bound is denoted by the constant DP_{uv}. The second constraint forces w_{uc} to zero if switch u is not mapped to controller c ($y_{uc} = 0$). κ is a large constant so the constraint is not binding when $y_{uc} = 1$. Additional constraints are needed to ensure each switch is assigned to exactly one controller, each controller is placed in exactly one location, two or more controllers will not be placed in the same location, and controller capacity will not be exceeded by the sum of demands of switches assigned to it. These constraints are straightforward to model and thus are omitted for brevity. Note that the objective and constraints of the problem are linear. Therefore, the problem can be solved using standard ILP solvers such as CPLEX. These solvers rely on branch and bound optimization techniques and can return the optimal solution for small and medium scale problem instances. However, they do not scale for problem instances with hundreds or thousands of variables. The evaluation in [45] considered small networks of sizes 10, 27, and 40 nodes for which the optimal solution can be found in reasonable time. The authors also presented a mechanism for dynamically assigning a switch to a back-up controller when all the control paths to its primary controller (according to the y_{uc} variable) have failed. They argued that heuristic rules of selecting as backup the closest controller instance or the controller with the highest residual capacity can work well in practice. The overall primary and backup controller placement method is referred to as "Survivor" and is one of the most extensively cited papers in the literature.

The authors in [46, 47] were among the first to recognize that there are multiple metrics and tradeoff considerations when it comes to optimizing the reliability in SDN. Therefore, a multi-objective rather than single-objective optimization approach is more suitable. They described four reliability objectives: (i) the controller-switch latency and how this increases when controller failures trigger reassignment of switches to backup controllers; (ii) the number of controller-less switches that cannot reach any controller due to link and node failures causing network fragmentation; (iii) load imbalance among controllers when switches are re-assigned from failed to other controllers; and (iv) the inter-controller latency that is conflicting with the controller-switch latency but is important for in-time synchronization among controllers to maintain a consistent global state. For the controller-switch latency, they considered as metric the maximum (wost-case) latency among all controller-switch pairs given by:

$$\pi^{\text{max latency}} = \max_{v \in V} \min_{c \in C} d_{vc},\tag{2.22}$$

where d_{vc} indicates the shortest path distance between switch v (in set V) and controller c (in set C). Here, the assumption is that each switch is always associated with the closest controller without considering the controller capacities. The interesting issue is how latency is affected when up to $k - 1$ out of the k controllers fail, i.e., only 1 controller remains active. In the worst case, the surviving controller will be the most distant controller over all switches, and hence the latency will increase to:

$$\pi^{\text{max latency}}_{\text{after failures}} = \max_{v \in V} \max_{c \in C} d_{vc}.\tag{2.23}$$

For the second objective, a switch is considered controller-less if it is still part of a working sub-topology (consisting of at least one more switch) but cannot reach any controller. Note that switches that are completely cut-off from all other switches are not considered controller-less since they have no neighbor to communicate with. To model this objective, we need to consider all the possible scenarios of failure of links and nodes which are of exponential number. For each such scenario s (in set S), we denote by $e^s_{vc} = 1$ if switch v cannot reach controller c, otherwise 0. Then, we define:

$$\pi^{\text{controller-less}}_{\text{after failures}} = \max_{s \in S} \sum_{v \in V} \min_{c \in C} e^s_{vc}.\tag{2.24}$$

In the above expression, the max term indicates the worst-case scenario. Then, for each scenario, each switch v is counted as controller-less if it is disconnected from all controllers so the min term takes the value 1. An interesting question to ask here is what is the minimum number of controllers and their placement to eliminate the occurrence of controller-less switches for up to a fixed number of edge and switch failures. This question is explored numerically for a small network where it is shown that eight controllers are needed to handle all failures of up to two switches or edges. Another finding is that a price is paid in terms of increased latencies to achieve reliability against controller-less switches (up to eight times).

For the load imbalance objective, [46, 47] define as metric the difference between the number of switches assigned to the controller with the most switches and the number of switches assigned to the controller with the fewest switches. The assumption is that a switch will always assign to its closest (non-failed) controller. With this rule, for each failure scenario s, we denote with n_c^s the number of switches assigned to controller c. The load imbalance metric is given by:

$$\pi_{\text{after failures}}^{\text{imbalance}} = \max_{s \in S} \left(\max_{c \in C} n_c^s - \min_{c \in C} n_c^s \right). \tag{2.25}$$

Again a price is paid in increased latencies when load balancing the controllers where the best placement according to one metric can significantly worsen the other metric.

Finally, the metric of inter-controller latency is briefly discussed in these papers. The argument is that to minimize this metric we need to place controllers as close as possible to one another and so adopting a more compact placement. This however increases the controller-switch latencies and the two metrics are conflicting. While this tradeoff is discussed, the impact of failures on inter-controller latencies is not analyzed and no mathematical formulas are given like in the three metrics above. However, the analysis should be similar to the controller-switch latency analysis.

To balance the four objectives above, [46] argues that an efficient method is to compute all possible solutions evaluated by all objectives and then pick one of the *Pareto-optimal solutions*. A Pareto optimal solution refers to a solution around which there is no way of improving any objective without degrading at least one other objective. In other words, if there is a way to improve one objective without hurting the other objectives, a Pareto optimal solution will find it. This approach can work in small- and medium-sized networks where optimal solutions can be found by brute force exhaustive search methods but not in larger networks. To fill this gap, the journal version of this work in [46] proposed a simulated annealing heuristic that requires much less time to run.

The work in [48] considered one of the four objectives mentioned above; the latency increase after failures. It concentrated on the scenario that up to $K - 1$ of the controllers can fail. To avoid disconnections due to controller failures, each switch maintains a list of K reference controllers. The switch will reassign to the next reference controller when a controller fails. We note that an implicit assumption here is that the switch should be constantly aware of the status of its K assigned controllers to reassign to the highest ranked available one at each time. This will increase the complexity of network management. The assignment of reference controllers to a switch must be so that the first reference controller is the closest to the switch, the second reference controller is the second closest, and so on. This way latency is kept to the minimum level possible. Since in the worst case $K - 1$ controller failures will happen, the objective is to minimize the maximum latency from any switch to its Kth reference controller, given by the equation:

$$\min \left\{ \max_{i \in V} \sum_{j \in V} y_{ij}^K \right\}. \tag{2.26}$$

Here, we slightly changed again the definition of the switch-controller association variable to follow the model in [48]. Specifically, $y_{ij}^k \in \{0, 1\}$ indicates whether the kth reference controller to switch i is j. The problem formulation includes a number of linear constraints that limit the number of placed controllers, couple the placement and the association decisions, restrict the association decisions according to the list of reference controllers, and the controller capacity constraints. The detailed description of these constraints can be found in [48] and is omitted here for brevity. Since the objective and all the constraints are linear and the variables are integers, the problem is an ILP and can be optimally solved for small and medium sizes using an ILP solver. The follow-up work from the same authors in [49] improved the model so that the switches are not required to be aware of the status of their K associated controllers. The idea is that a switch will forward its control request always to the switch where its first reference controller is located. The latter switch is the only one that is aware of the failure or not of that controller and will re-direct the request to the local controller or to the second reference controller in case of failure, and so on. A drawback of this approach, however, is that the latency is increased every time the request is redirected. Besides, an implicit assumption is that the controller might fail but not the switch where the controller is placed. Therefore, this work cannot capture all types of failures in the network. A simulated annealing heuristic algorithm is proposed to solve the problem in a computation efficient manner.

The work in [50] proposed associating each switch with multiple controllers so as to improve reliability (like [41, 42]). They formulated the problem of controller placement and switch association with the goal of minimizing the number of controllers while satisfying a number constraints; a lower-bound constraint on the number of controllers associated with each switch, a maximum latency constraint between each switch and its associated controller, another maximum latency constraint between each pair of controllers, a controller capacity constraint, and a constraint that couples the controller placement and switch association decisions. We note that the maximum latency constraint between controller pairs is not a linear function, but given by:

$$x_u x_v d_{uv} \leq \Delta^{inter}, \tag{2.27}$$

where Δ^{inter} is a constant and d_{uv} is the latency between switches u, v where controllers are placed ($x_u = x_v = 1$). Yet, this constraint can be easily linearized using the McCormick envelopes by defining a new variable $x'_{uv} = x_u x_v$ and subsequently replacing the constraint in Equation (2.27) with the following set of constraints:

$$x'_{uv} d_{uv} \leq \Delta^{inter} \tag{2.28}$$
$$x'_{uv} \leq x_u \tag{2.29}$$
$$x'_{uv} \leq x_v \tag{2.30}$$
$$x'_{uv} \geq x_u + x_v - 1 \tag{2.31}$$
$$x'_{uv} \in \{0, 1\}. \tag{2.32}$$

By linearizing the constraints, the problem can be solved as an ILP for small and medium network sizes. To handle larger problem instances, [50] proposed a heuristic algorithm that is based on the clique concept in graph theory. The key observation is that each switch and its assigned controllers can be regarded as a clique since the switch is directly connected to all the controllers and the controllers are directly connected themselves. Then, an iterative process that places controllers and assigns switches corresponding to different maximal cliques can be adopted to find a heuristic solution.

Control Path Optimization

All of the work described above assumed that the control paths (i.e., paths between controllers and their associated switches as well as inter-controller paths) are given (e.g., [37, 38]) or used simple heuristic policies for determining these paths like picking always the shortest paths (e.g., [46, 47]) or computing a set of disjoint paths by solving a max flow problem (e.g., [41, 42]). Given the availability of these paths, they were able to model and optimize the controller placement and switch association decisions. However, the selection of control paths can be explicitly optimized to improve further the problem objectives. In the sequel, we describe three studies that considered such more general version of the problem.

The work in [51] proposed a mathematical formulation that explicitly optimizes the control paths between the switches and their associated controllers (but not the inter-controller paths). It considered two cases: (i) every switch must be connected to its associated controller through two disjoint paths (one primary and one backup) and (ii) every switch must be connected to two different controllers (one primary and one backup) over two disjoint paths. Clearly, the second case is more reliable to failures since it can also handle the failure of one of the controllers. To model these problems, we need a variable $I^p_{u,vw}$ and $I^b_{u,vw}$ to indicate whether edge (v, w) belongs to the primary and backup control path of switch u, respectively. These variables are in addition to the controller placement and switch association variables (x_u, y_{uv}). The goal is to find a controller placement that provides primary and backup control paths of the shortest possible lengths:

$$\min \sum_{u \in V} \sum_{(v,w) \in E} (I^p_{u,vw} + I^b_{u,vw}) d_{vw}. \tag{2.33}$$

Flow conservation law has to hold for both primary and backup paths:

$$\sum_{(v,w) \in E} I^p_{u,vw} - \sum_{(w,j) \in E} I^p_{u,wj} = y_{uw} - 1_{\{u=w\}} \tag{2.34}$$

$$\sum_{(v,w) \in E} I^b_{u,vw} - \sum_{(w,j) \in E} I^b_{u,wj} = y_{uw} - 1_{\{u=w\}}, \tag{2.35}$$

where $1_{\{u=w\}}$ is the indicator function; it is equal to 1 if the condition to the subscript is true, otherwise 0. Here, the first equation ensures that a flow belonging to a primary control path of switch u and entering switch w has to leave the switch w unless (a) the switch w is the controller associated with u ($y_{uw} = 1$) or (b) the switch w is the switch u ($u = w$) or (c) or both. The same

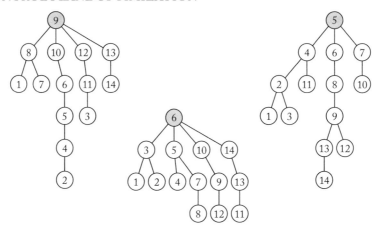

Figure 2.4: Overlay tree networks for communication between the controller at the root solid node and the switches in [52]. The number at the end of each branch is the delay in it.

for the backup control path in the second equation. The problem belongs to the class of ILP and can be solved with ILP solvers.

The work in [52] proposed another method for reliable controller placement referred by k-critical, analogously to the k-median and k-center methods that have been extensively used to optimize latency [24]. Here, the main idea is that each controller will manage the switches associated with it by forming an overlay tree of control paths. We note that this overlay tree concept is similar to what we previously described when we presented the work in [39]. However, [39] did not optimize the structure of the tree.

The intuition here is that along with the controller placement we need to design trees that are robust so that the communication remains in spite of network failures, thus minimizing any data loss. Consider for example the three trees in Figure 2.4 found by the k-center, k-critical, and k-median methods when applied to a network in [52]. If we consider the worst case when a failure occurs in the network, which is when the adjacent link to the controller with highest number of nodes in the branch fails, we found that tree (a) loses information from six nodes, tree (b) from four nodes, and tree (c) from seven nodes. So, as we can see in Figure 2.4, the shortest and load balanced paths reduce the loss probability and bottlenecks in the tree. Motivated by this, the k-critical method picks controllers in a way that the branches of the resulting trees will have a maximum delay limited by a threshold value.

The work in [53] is the last we discuss among the reliable controller placement works. It focuses on the construction of reliable control paths for the inter-controller communication (rather than the switch-controller communication). To save communication overheads, a multicast model based on a spanning tree between the controllers can be used instead of a unicast model. Then, the question is how to construct such a spanning tree which is more complex than

Table 2.3: State-of-the-art works on cost-centric controller placement

Ref.	Objective	Controller Count Opt.	Controller Capacity	Controller-Switch Assoc.	Method
[54, 55]	min controller installation, controller-switch and inter-controller linking costs	✓	✓	✓	ILP
[56]	min controller cost, basic costs, controller-switch and inter-controller link delay costs	✗	✓	✓	Greedy network partition
[57]	min energy cost	✗	✓	✓	Genetic algorithm
[58]	min controller-switch and inter-controller traffic	✓	✗	✓	Facility-location heuristic
[59, 60]	min controller-switch traffic, inter-controller traffic and controller switch latency	✓	✗	✓	Submodular optimization and facility location approximation

simply computing unicast shortest paths for the different controller pairs. Therefore, a problem formulation is proposed in which along with the classic controller placement (x_u) and switch assignment (y_{uv}) variables, extra variable for all links are introduced to indicate whether that link is part of the spanning tree or not. Similar to [51] discussed above, a flow-based formulation is adopted to determine the links of the tree. Additional effort is made to model the scenario that a single link in this tree fails and a failover mechanism is triggered to connect the two endpoints of this link using a shortest path. This failover, however, will increase the delay of the tree. Therefore, the objective is to minimize the tree delay for both the cases without any failure and for the case with a single arbitrary link failure. Since both the objective and constraint are linear the problem can be solved as an ILP. The detailed formulation is omitted for brevity.

2.1.4   COST-CENTRIC METHODS

Another objective of the problem is the cost associated with the deployment and operation of controllers. These are known as the capital expenditures (CAPEX) and operating expenditures (OPEX). The CAPEX cost has to be paid once and includes the monetary cost of purchasing the machines that host the controllers. It also includes the cost to deploy the controllers and connect them with each other and to the switches in the network. We note that the connection cost will be lower if the controllers are located in the same locations as the existing switches in the network. In this case, the existing switch-to-switch paths can be used for establishing controller-to-controller and controller-to-switch connections rather than creating new (costly) links (out-of-band approach). The OPEX cost includes the bandwidth and energy consumption due to the inter-controller and controller-switch communication.

The work in [54] proposed a model aiming to calculate the optimal number, location, and type of controllers that minimizes the overall cost. The latter includes the cost of controller installation, linking of controllers to switches, and interlinking of controllers. At the same time, it needs to satisfy constraints like controller and link capacity is not exceeded. The work in [55] extended that work with the objective of minimizing the cost of expanding an existing SDN network with constraints similar to that of their previous work. A limitation of both these works is that they require solving an ILP, using a solver like CPLEX, and therefore they will work only for small-scale networks.

The work in [56] clusters the controllers into sets where each sets involves some basic costs like warehouse rent, power supply cost, and man power. The controller placement problem is formulated as an ILP that aims to minimize the controller cost, basic costs, switch-to-controller, and controller-to-controller costs. The proposed solution partitions the network topology into several SDN domains and places a controller set in each domain. A greedy algorithm is proposed to partition the network which exploits the property of minimum nonzero eigenvalue of Laplacian matrix to cluster closer nodes to a SDN domain by considering the minimum non-zero eigenvalue.

The work in [57] argued that the controller placement also affects energy consumption. They formulated the problem of placing a given number of controllers, associating switches with them, and selecting the network paths for this association. When a link is selected by at least one network path it has to be active and the respective energy cost has to be paid. The solution has to satisfy several constraints such as the latency of the paths and the capacity of the controllers. However, the energy cost associated with the inter-controller paths was not considered. The problem is formulated as an ILP that can be optimally solved for small-scale networks. For large-scale networks, a heuristic genetic algorithm is proposed where a chromosome represents a placement solution of the controllers.

The work in [58] aimed to minimize both the controller-switch communication cost as well as the inter-controller communication cost. These costs can capture both the bandwidth consumption and the latency of the respective communication. The costs are defined as constants which are equal to the number of hops between the respective pairs of nodes (controllers or switches). The authors formulate the problem as an ILP. Due to the high computation complexity of ILP, the authors also propose a greedy approximation algorithm. The approximation solution decomposes the initial problem as several classical facility location problems and selects the minimum cost among them as the final result. The heuristic approach sorts the nodes in decreasing order of their degree and returns a placement such that the synchronization cost and communication cost is below a threshold.

The work [59, 60] performed experiments with two real controller platforms, OpenDaylight and ONOS, to measure the traffic overheads between controllers and switches and drive the design of a model. The experiments demonstrated that the inter-controller traffic for state synchronization increases roughly linearly with the load of the controllers, while the pattern is different between the two controllers platforms due to the different consistency protocols used. For the ONOS platform, a heuristic algorithm was proposed using the theory of submodular function optimization which allows for proving approximation guarantees of the solution to the control traffic cost minimization problem. For the OpenDaylight platform, a facility location inspired algorithm was proposed.

It becomes clear from the above that the modeling methodologies of the controller placement problem in literature vary largely with the particular application scenario under consideration and the time of the publication. Different authors model in different ways parameters such as the controller location, type, capacity, and count as well as inter-controller traffic for state synchronization. Besides, as it happens with many other problems, researchers incrementally consider more and more twists of the problem and derive more and more complex formulations to address simplifications and limitations of the early works. For instance, the first works in the area (e.g., [24]) considered simplified versions of the problem were the number of controllers is assumed to be known and the capacity of the controllers is not an issue. There, only the location of the controllers needs to be optimized to improve propagation latency. Subsequent works relaxed these assumptions and proposed more complex models, e.g., with hard constraints on the controller capacities and more elaborate latency models, e.g., using queuing theory to model the processing latency. As the models were becoming more complex different types of controllers were also taken into consideration, with different characteristics and costs.

2.1.5   ADAPTIVE METHODS

In the methods described so far, the controller placement as well as the association between controllers and switches (**x** and **y** variables in the optimization problems) were fixed based on some initial view of the network and objectives. However, as traffic conditions change dynamically, the initial decisions we make in the control plane may no longer fulfill the intended objectives. For instance, based on real measurements on real data center networks, the peak-to-median ratio of flow arrival rates during the same day can be almost 1–2 orders of magnitude [66]. Such traffic variations can cause imbalance among the controllers, i.e., some overloaded and some underutilized. An overloaded controller will response to switch requests with increased latency, deteriorating the quality of user experience. To address these issues, we can use methods that vary the number of controllers and their switch associations over time to adapt to the changing network conditions. We begin with describing the methods that adapt only the association of switches with controllers for a fixed controller placement. Following that, we discuss more advanced methods that adapt also the number and location of controllers.

Adaptive Switch Association

Re-associating or migrating a switch from one controller to another can cause disruption to ongoing flows which can severely impact the various network applications. OpenFlow does not natively provide any support for disruption-free switch migration. A disruption-free migration protocol is therefore needed to ensure there is no impact on traffic flows. Such protocol should guarantee liveness and safety properties, as described below.

- **Liveness.** At least one controller is active for a switch at all times. For each switch request, the controller that issues the request remains active until the request is processed.

- **Safety.** Exactly one controller processes each switch request. Duplicate processing could result in duplicate entries in the flow table of the switch.

The work in [67] proposed a switch migration protocol that guarantees the above properties. To perform the migration of a switch from the first to the second controller, the protocol requires four phases which exploit OpenFlow's "equal mode" semantics. Initially, the second controller changes its role for that switch to "equal" that allows it to receive control messages (e.g., PacketIn). Then, the first controller sends to the switch a dummy flow setup command and shortly after a flow deletion command. The key point is that the switch will reply to the latter command by contacting both controllers. This will be the triggering event that from now on the second controller is the responsible one for that switch. In the next phases, the protocol ensures that the first switch will handle any pending requests received before the triggering event, and finally the second controller will change its role to "master" which declares the end of the migration process.

Motivated by disruption-free migration protocols like the above, researchers have looked at the switch migration problem from a modeling and optimization point of view where the goal is usually to load balance the controllers or a variant of this metric. To this effort, tools from various fields have been employed including Network Utility Maximization (NUM) frameworks, Markov chain approximations, game theory, graph theory, and stochastic control theory. Besides, the problem has been studied at different *granularities* of decision making. While early work focused on migrating a switch in its entirety from an overloaded controller to a less loaded one, later work makes decisions on a per-flow basis. This is achieved by associating a switch with multiple controllers while at the same time different flows emanating from that switch can migrate to a different one of these controllers. Table 2.4 lists the main body of work in this area classified according to the objective, granularity of decisions, and solution method, as well as based on whether they operate in a centralized or decentralized manner, and other modeling assumptions (e.g., whether an explicit model of migration cost is adopted or not).

The work in [68] formulated the switch migration problem as a *Network Utility Maximization (NUM)* problem for a switch decision granularity. Instead of using integer variables $\{y_{uv}\}$ to model the switch association decisions, we can alternatively employ a *graph partitioning*

Table 2.4: State-of-the-art work on adaptive switch association

Ref.	Objective	Decision Granularity	Migration Cost	Operation	Method
[68]	max network utility	Switch	✗	Decentralized	NUM, Markov Approximation
[69]	max network utility	Switch	✗	Decentralized	Greedy network partition
[70]	min controller response time and control traffic overhead	Switch	✗	Decentralized	Stable matching
[71]	min blocking probability, max controller capacity utilization	Flow	✗	Centralized	Threshold-based policy
[72]	min cost function	Flow	✗	Centralized	Flow-request partition heuristic
[73]	min cost function	Flow	✗	Centralized	Cost-based greedy heuristic
[74]	min flow setup time	Flow	✗	Centralized	ILP
[75]	min variance of controller loads	Switch	✓	Centralized	Heuristic based on switches group
[76]	max migration efficiency	Switch	✓	Centralized	Greedy
[77]	max migration efficiency	Switch	✓	Centralized	Heuristic based on trigger factor
[78]	min max controller load	Switch	✗	Centralized	Heuristic based on clustering and greedy methods

problem formulation where the variables are sets of switches (rather than integer values). Specifically, for a given set of controllers $C = \{c_1, c_2, \ldots, c_k\}$, the question is how to partition the set of switches V in k-partitions $\{V_i\}_{i=1}^{k}$, with $V_i \subset V$, $V_i \cap V_j = \emptyset$, $i \neq j$. Here, V_i indicates the set of switches associated with controller c_i. Then, the utility of a controller c_i can be defined as an increasing and concave function of its load, such as:

$$U_i = \sum_{u \in V_i} \log(\lambda_u). \tag{2.36}$$

The objective of the NUM problem is to find a graph partitioning such that the total (over all controllers) utility is maximized:

$$\max \sum_{c_i \in C} U_i. \tag{2.37}$$

The rationale of picking such an objective function is that the utility of a controller should increase with the number of control requests it serves from a switch but the rate of increase should decrease so that other switches also receive service. Given the exponential number of possible partitions, the NUM problem is challenging to solve. Applying integer solvers to this problem would be time-consuming as the network size becomes larger. To reduce computation time, [68] proposed an approach that can be concurrently processed in a decentralized manner, separately at each controller with some limited only interaction with its neighbors. This approach relies on a Markov approximation framework where each possible switch partition $f = [V_1^f, V_2^f, \ldots, V_k^f] \in F$ is associated with a probability p_f representing the percentage of time this partition is selected. Then, an equivalent formulation is to find the probabilities so that:

$$\max_{p_f \geq 0} \sum_{f \in F} p_f \sum_{c_i \in C} \sum_{u \in V_i^f} \log(\lambda_u). \tag{2.38}$$

The next step of the approach is to approximate the optimal value of the above problem by using a log-sum-exp function that will allow us to find the optimal probability values $\{p_f^*\}$ in closed form. With this information at hand, we can define a continuous-time time-reversible ergodic Markov chain with stationary distribution equal to the above probability values. The key idea is to regard the states of the chain as switch partitions and the transitions from one state to another as migrations of switches among controllers. The last step is to design a decentralized process where each controller counts down a clock and waits for migrating a switch to another controller until clock terminates, where the switch migration probability follows the transition probability in the Markov chain.

The work in [69] proposed another decentralized switch migration method, this time based on *non-cooperative game theory*. The main idea is to initiate a new game each time a controller becomes overloaded. In this game, other nearby controllers with residual computation resources are the players, and the switches under the administration of the overloaded controller are the commodities. The players compete for the commodities, each player having a payoff function similar to the one in Equation (2.36). The game is played for each commodity (switch) in a sequential manner starting with the switches generating more load and being more far away from the controller that initiated the game. Among the competing players, the one with the largest difference in resource utilization will be selected for the switch migration. The game will continue to be played and switches will be migrated one by one until the controller is not overloaded any more.

The work in [70] modeled the switch association problem as a *stable matching problem with transfers* and proposed a two-phase algorithm. In the first phase, the problem is transformed into the classical college admissions problem. A switch's preference over controllers is based on the worst response time that the controller can provide, whereas a controller's preference is based on the control traffic overhead caused by the communication between them. Then, a stable matching which guarantees the worst-case response time for switches is obtained. This is used as the initial partition for the coalitional game in the second phase. In this scheme, switches participate in a Nash stable solution, where none of them has an incentive to change to a different controller that can lower its response time while not harming others' performance. The advantages of the two-phase algorithm are two-fold: first, stable matching is competitive in its outcome and efficiency. The algorithm generates a stable matching that can be easily implemented in a centralized manner with low time complexity, which is suitable for large scale problem instances. Second, the two phases are complementary. The solution of the stable matching phase serves as the input of the coalitional game and accelerates the convergence of the second phase while the coalitional game makes transfers to further improve response time.

The methods described so far achieve controller load balancing by migrating switches from overloaded to less loaded controllers. The work in [71] proposed a load balancing method using a *per-flow basis*. The idea is to keep the association between switches and controllers static. However, when a new flow request is generated at a switch and is sent to the associated controller, the latter, if overloaded, can migrate the flow request to another controller. By getting a second chance, it becomes more likely for the flow request to be served rather than being blocked due to controller overload. The authors provide an analysis of the probability of blockage in the traditional method (without flow migration permitted) and the proposed method. Specifically, they model the new and migrated flow request arrivals as Poisson distributions with known rates and then adopt a Markov chain model where each state refers to the number of pending requests at each controller. Therefore, state transitions happen as new or migrated flow requests arrive at the controller or previously arrived requests are processed. The probability of blockage as well as the controller capacity utilization are expressed as functions of the steady probability distribution of the Markov chain. Numerical results demonstrate the performance superiority with respect to the above two metrics of the proposed method compared to the traditional method. While the idea of flow request migration is interesting and the analysis based on the Markov chain is technically sound, this work has a major drawback. It does not consider the extra traffic overhead and latency due to the migration of the flow request from one controller to another.

The work in [72–74] proposed a different approach that allows each switch to be associated with multiple controllers. This can be realized with OpenFlow by defining different roles of controllers with respect to the same switch. To handle short-term traffic variations, these works propose to dynamically distribute fractions of the flow requests of the switches to the multiple different associated controllers. The switch association remains relatively static and will change only on a slow timescale. The rationale behind this approach is that frequent re-association of switches with other controllers is undesirable due to the overhead required for transferring state information from the controller previously handling the flows at the switch to the new controller. On the other hand, re-distribution of flow fractions among controllers is more practical. In a nutshell, the two approaches—switch re-association and flow re-distribution—should be regarded as complementary, targeting the same goals albeit running at different timescales.

In more detail, the work in [72] proposed BalanceFlow, a load-balancing controller architecture. The key feature of this architecture is the super controller, a controller with special role that is responsible for balancing the load of all other controllers. To do this, the super controller collects flow-requests information from all controllers. When load imbalance is detected, the super controller re-distributes the flow requests across controllers and installs on switches the respective allocation rules. The objective is to re-distribute the flows in a way that a cost function is minimized. This cost function is defined so as to favor the association of switches with controllers that are less loaded and with small propagation latency. A similar load-balancing architecture but for hierarchical SDN controller deployments was proposed in [73]. Both use heuristic algorithms to minimize a cost function. A limitation of both works is that the distribution of flow requests is done by a rather coarse-grained approach based on source-destination switch pairs of the flows. Another limitation is that the switch is assumed to have enough processing capability to locally do by itself the flow request distribution to the different controllers it is associated with. A more fine-grained approach where flow requests are considered at the individual flow level instead of aggregating flows based on source-destination switch pairs was proposed in [74]. The problem is formulated as an ILP where the objective is to minimize the flow setup time. This is defined as the sum of response time at the controller (or in other words the processing latency) and the two-way propagation latency between the switch and the controller. The first term captures the need for load balancing the controllers since an overloaded controller will have higher response time. Another interesting aspect is a resilience type of constraint which guarantees that a certain percentage of new flows at a switch will not be affected by a single controller failure. The ILP will be solved in a semi-online manner; every time significant increase in the response time is detected.

The work in [75] considers another aspect of the problem, *load oscillation*. This term describes the undesirable event that switch migration while offloading one controller causes the shifted load to overwhelm the second controller. To address this problem, a heuristic method is proposed that selects a group of switches whose load is the nearest to the difference of two loads—the load of the overwhelmed controller and network average. This makes the load of the overwhelmed controller to be closer to the network average after the migration is completed. To reduce the control plane latency, the algorithm chooses the farthest switch's group with the overwhelmed controller to migrate when there are multiple groups whose loads are equally close to the target value. To avoid load oscillation among controllers, the load of the second controller is made to be near to the network average load as much as possible. The migration overhead is also considered in the selection.

With the exception of [75], the rest of the work we described above did not consider the cost of migration. These works focused on utilizing/load balancing the available controllers without taking into account the traffic overheads and network instability that such migration might bring. The work in [76] explicitly modeled the migration cost. This mainly has two components: (1) the increase in load cost and (2) the message exchanging cost. This work defined a new metric referred to as migration efficiency which is the ratio of load balance variation to migration cost. Therefore, the purpose of the migration efficiency metric is to select a controller as the destination controller for load shifting based on the principle of migration efficiency maximization. A greedy algorithm is proposed that selects the migrations of switches that lead to the highest increase in migration efficiency.

The work in [77] proposed Efficiency-Aware Switch Migration (EASM), a dynamic switch migration framework whose objectives are similar to that in [76]. However, unlike [76] the migration efficiency is the ratio of load balancing rate to switch migration cost. The migration cost is computed in terms of the average size of PacketIn messages and migration hops between the switch and controller. The authors introduce a load difference matrix and trigger factor which is a measure of the load variance of the controllers. If the trigger factor exceeds a threshold the switch migration process is invoked.

The last work in Table 2.4 is [78]. We discuss this last and in more detail because it has one major difference compared to all the others. Specifically, all the other work assumed that a PacketIn message generated at a switch will be handled by one of the controllers associated with that switch and computation load will be incurred only by that controller. However, if the PacketIn message corresponds to a flow whose destination is a switch outside the domain of that controller then the same flow request will trigger additional PacketIn messages when the flow exits one domain and enters another domain. To capture this, a more sophisticated control load model that takes into account the network paths of the flows is required. Such a model was proposed in [78] as summarized below.

The model considers an SDN network composed by V switches and C controllers. The traffic can change over time in an unpredictable manner. Yet, we can monitor the current traffic load during runtime. The notations f_{oS_i}, $f_{S_i o}$, and $f_{S_i S_j}$ indicate the current arrival rate of new flows that enter the SDN network from switch S_i, leave the SDN network from switch S_i and traverse the link between the two connected switches S_i and S_j, respectively. These new flows are responsible for the computation load at the controllers. Specifically, the computation load of a controller C_m is composed of three components: (a) the path computation load $\mathcal{K}_{f_o S_i}$ of the flows arriving from outside the SDN network to a switch S_i managed by controller C_m; (b) the path computation load $\mathcal{K}_{f_{S_j} S_i}$ of the flows arriving at switch S_i from S_j, a switch controlled by another SDN controller $C_n \neq C_m$; and (c) the flow rule installation load at each switch S_i managed by controller C_m. The latter of the three components can be modeled as:

$$\sum_{S_j \in V} \mathcal{G}(f_{S_i S_j}) + \mathcal{G}(f_{S_i o}), \tag{2.39}$$

where function $\mathcal{G}$ maps the flow arrival rate at S_i to the computational load at the SDN controller needed for rules installation. Note that the above equation expresses the amount of flows that are traversing S_i going to any other switch S_j or out of the SDN network.

Having defined an appropriate model for translating flow arrival dynamics at the switches to computational loads at the controllers, we can now formulate the switch assignment problem. Similar to the work in [68], we express this problem as a *Graph Partitioning Problem* where the question is to partition the set of switches V in $|C|$-partitions $\{V_m\}_{m=1}^{|C|}$, with $V_m \subset V$ and $V_m \cap V_n = \emptyset$ for each $m \neq n$. For a given partition, the computation load at controller C_m is given by:

$$LC_m = \sum_{S_i \in V_m} \mathcal{K}_{f_o S_i} + \sum_{S_j \notin V_m, S_i \in V_m} \mathcal{K}_{f_{S_j} S_i} + \sum_{S_i \in V_m, S_j \in V} \mathcal{G}(f_{S_i S_j}) + \sum_{S_i \in V_m} \mathcal{G}(f_{S_i o}). \tag{2.40}$$

The goal of the optimization problem is to find the graph partitioning that minimizes the maximum controller load:

$$\min_{V_1, \ldots, V_{|C|}} \max_{C_m \in C} LC_m \tag{2.41}$$

$$\text{subject to: } V_m \cap V_n = \emptyset, m \neq n \tag{2.42}$$

$$\bigcup V_m = V. \tag{2.43}$$

To get some intuition about the problem, consider the example network in Figure 2.5. The partition in Figure 2.5a assigns half of the switches to each of the two controllers. However, two out of the three sets of new flow arrivals traverse switches of both domains. Thus, the path computation load for these flows has to be paid twice, one time per controller. A more efficient partition is the one in Figure 2.5b where each flow is completely managed by a single controller. Thus, the path computation is paid once for each flow.

As shown in [78], the graph partitioning problem is NP-Hard as there exist an exponential number of possible partitions to be examined in order to find the optimal one. An option is to use a heuristic algorithm. One such heuristic algorithm was proposed in [78]. Here, the main idea is that switches that are highly connected, with high volume of traffic among them, should be

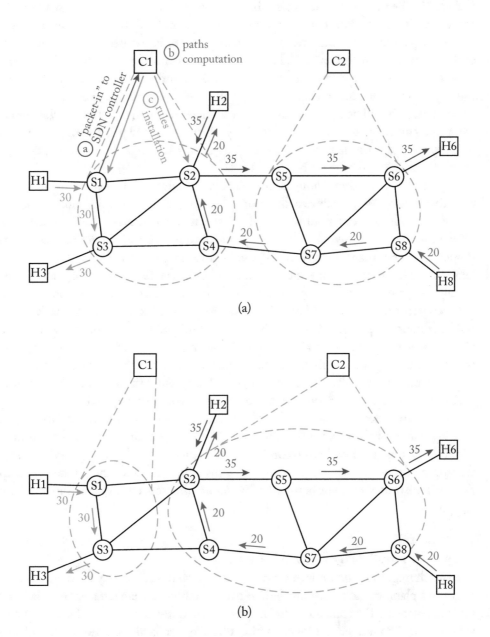

Figure 2.5: (a) SDN controller load imbalance scenario. (b) Controller load balance is improved after switch migrations [78].

included in the same partition. This is intuitive since if these switches are managed by different controllers, then the path computation cost for the traffic flows traversing these switches will have to be paid multiple times causing extra load to multiple controllers. Therefore, the first step of the heuristic algorithm is to find *clusters* of highly connected switches. The algorithm should be able to update the clusters, e.g., expand or shrink, during run time to adapt to changes of the traffic patterns. The second step of the algorithm will be to assign these clusters of switches to controllers. This will be done in a dynamic and incremental manner where clusters are migrated from one controller to another. Since the problem is NP-Hard, the migration will be based on simple greedy criteria such as for each cluster checking if there are alternative controllers to migrate that can lead to more balanced loads.

Migrating switches among controllers dynamically based on the controllers' load can balance their loads so as to relieve congestion. However, during the migration, some switches may not be able to handle new connections in a timely manner. This is called *migration blackout* [67]. While the four-phase switch migration protocol proposed in [67] guarantees some important properties (liveness and safety) as discussed above, migration backout still constitutes a problem. There are two main ways to tackle this problem. The first is by incorporating a penalty into the algorithm each time a migration of a switch takes place, and so seeking to find a tradeoff between performance of load balancing and cost of migration. Here the blackout problem will not be completely prevented, although it will be mitigated. The second way aims to completely alleviate the blackout problem by temporarily steering newly arriving flows away from switches that are to be migrated so that their flow setup will not be affected by the migration. In order to ensure that there is always an alternative path bypassing the switches to be migrated, the algorithm should add a new constraint to the selection of migrating switches—the (hypothetical) removal of the selected switches should not break network connectivity. Note that this traffic steering may slightly increase the hop count of the forwarding path of some flows. But compared to the possibly hundreds of milliseconds of migration blackout period, the slightly larger forwarding delay due to a longer path can be insignificant. Once the migration completes, new flows will be routed on their best paths. Therefore, the impact of this method on latency is only temporary.

Adaptive Controller Number and Location

The switch migration methods in the previous section paved the way for methods more adaptive to control plane traffic variations. While these methods have been shown to be quite effective for load balancing the controllers and improving their response time to flow setup requests, they assume that the number and location of controllers are fixed, i.e., they do not change to adapt to traffic variations. Nevertheless, some of the controllers may be lightly loaded at some times and could be turned off for energy and cost savings while migrating the switches under their administration to other controllers that are active. The implementation of such dynamic methods is facilitated by network function virtualization (NFV) technologies in which controllers can be

Table 2.5: State-of-the-art work on adaptive controller number and location

Ref.	Objective	Controller Architecture	Adaptation Cost	Operation	Method
[79]	min statistic collection, flow setup, synchronization and switch reassignment costs	Flat	✓	Centralized	Greedy, Simulated Annealing
[80]	min end-to-end flow setup time	Flat	✗	Centralized	ILP
[81]	min latency, # active controllers, max controller utilization	Flat	✗	Centralized	Location search algorithm based on network partition
[82]	min controller cost	Flat	✗	Decentralized	Non-Zero-Sum Game
[83]	min max controller load	Hierarchical	✗	Decentralized	Stochastic subgradient descent

realized as virtual machines that can be dynamically instantiated and migrated from one server to another in an automated way and with relative ease. Table 2.5 lists the state-of-the art methods in this area. These methods are classified based on the controller architecture (flat or hierarchical), whether they take into account the cost of adapting the control plane decisions, as well as based on their centralized/decentralized operation.

The pioneering work in [79] formulated the adaptive controller placement problem in WANs. Here, the controllers are running on servers located at designated locations throughout the network. The idea is that a controller is regarded as active if it has at least one switch associated with it otherwise it is considered inactive. Inactive controllers can only receive HELLO messages from newly associated switches, and they consume very low CPU power compared to the active controllers. The method should periodically collect flow statistics from the network, and decide how to adapt the association of switches with controllers which will directly determine which controllers are active and which are not. To make the most effective adaptation decisions, the method considers four types of costs: statistics collection cost, flow setup cost, synchronization cost, and switch re-association cost. Intuitively, many controllers should be activated across the network to reduce the cost of statistics collection by the controllers. This policy would also favor the flow setup cost although this metric depends also on the way the switches are associated with controllers and on whether more than one controller needs to be queried to

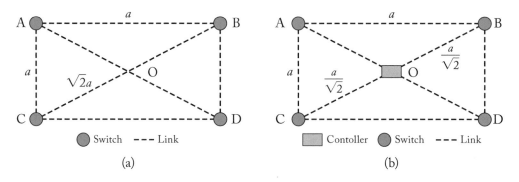

Figure 2.6: (a) An SDN network with four switches and (b) the impact of the controller placement on the latency of SDN [81].

establish the routing path of the flow (in case of inter-domain flow). On the other hand, the synchronization cost would be lower if fewer controllers are activated close one to another. Finally, the switch re-assignment cost captures the delay and traffic overheads of re-association process and is modeled in the problem formulation as an XOR function of the previous and current decision. The problem is modeled as an ILP that can be solved optimally for small networks. For larger networks, two heuristics are proposed, one greedy approach based on the knapsack problem, and another simulated annealing based meta-heuristic approach.

The work in [80] extended [79] by considering a more detailed model of the flow setup time. The authors define the end-to-end flow setup time as the time taken to forward the first packet of a flow from source to destination. For flows with a few packets, like DNS requests, this metric is a good indicator of network responsiveness and user experience. This metric consists of three components. These components include the flow setup time when the switches are in the same SDN domain, when the switches are in different control domains, and the forwarding latency between the source and destination nodes. The objective is to minimize the average flow setup time. The problem is formulated as an ILP where either the placement of the controllers or the associations between switches and controllers can change or both the placement and associations can change synchronously. The problem is solved using generic solvers like CPLEX. The adaptation cost, however, is not explicitly modeled.

The work in [81] investigated another interesting twist of the problem. While most of the other works in the literature considered fixed locations as candidates for controller placement, typically a subset of the switches, this work allowed the arbitrary placement of controllers in the entire geographical region where the SDN network resides. This approach provides more degrees of freedom in choosing the controller locations which can help in reducing the latency or the number of controllers required to reach a target QoS.

Figure 2.6 presents an illustrative example where four identical fully connected switches (A, B, C, and D) are considered at four corners of a square region. According to the common approach used by most works, the optimal location of the controller is one of four locations of switches (A, B, C, and D). For any of these locations the maximum latency, between the switches and the controller, is $\sqrt{2}a$, shown in Figure 2.6a. But, if the controller is placed in the center of the square region, shown in Figure 2.6b, then the maximum latency becomes $\frac{a}{\sqrt{2}}$. This controller placement halves the latency compared to the placement in the first figure. Thus, instead of the fixed locations, if the entire region containing the switch's locations is searched, the location found further decreases the latency.

Consider the same example from Figure 2.6a with the latency bound of $\frac{a}{\sqrt{2}}$. Now, according to the common controller placement approach, the controller placement is limited to the switch locations. Since the minimum latency between two consecutive switches is a, four controllers are required at four switch locations, to maintain the latency constraint. On the contrary, if the controller is placed in the center of the square region, shown in Figure 2.6b, only one controller is required to maintain the latency constraint. It reduces the number of controllers by four times! Thus, instead of the fixed locations, if the entire region containing the switches' locations is considered, a lower number of controllers is used.

The authors propose two search techniques to find the placement of controllers. In the open search technique, given the location of the switches the entire region is searched to find the optimal placement of controllers. Although the open search technique results in maximum utilization of the controllers, the approach is not practical as controllers cannot be placed at any location. Thus, a restricted search problem is introduced, where placement of the controllers is restricted to a set of selected locations. The authors find the location of the controllers using a location searching algorithm (LSA). The algorithm partitions the network into sub-regions using latency bounds and it uses a heuristic to find the smallest enclosing circle, given the location of switches. The heuristic is based on Welzl's algorithm with the controller being placed at the center of the circle.

We should emphasize at this point that while the controller placement in the entire geographical region can improve the performance compared to the more restrictive placement at the switches, it requires the extra cost of connecting with new links with the placed controllers in the network. This extra cost needs to be carefully considered in the model and be justified by the potential benefits in performance.

The methods discussed so far are centralized and may limit scalability for large-scale SDN systems. The work in [82] proposed a decentralized controller addition/deletion method where each controller makes its own decisions by solving an individual optimization problem. Since these individual problems are non-strictly competitive, the overall problem is solved as a non-zero sum game. Further, the authors reason that to make the approach adaptive, the equations need to be solved iteratively. However, they do not mention the frequency of the iterations. Besides, the adaptation cost is not explicitly modeled.

We discuss last the work in [83]. This work presents a more elaborate model where, in addition to adaptation of controller placement and switch association, the controllers can adapt

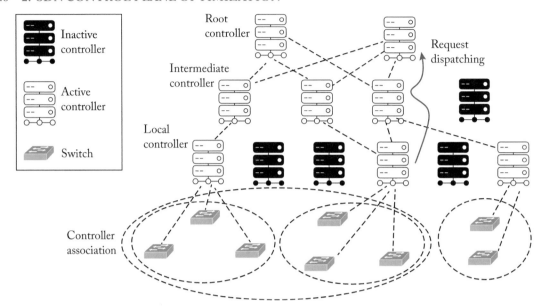

Figure 2.7: An illustrative example of an hierarchical control plane with adaptive controller activation, switch association, request processing, and dispatching [83].

the computation power they dedicate to process requests or even dispatch some of these requests to other controllers. Besides, the problem is formulated as an online optimization framework in which the decisions made in a time instance affect the future ones as we explain in the sequel. This is a completely different approach from the works we discussed so far where the decisions were adapted periodically, in a heuristic and myopic manner.

The problem is modeled for the case that the architecture of controllers is *hierarchical* (rather than flat), as shown in Figure 2.7. In the hierarchical architecture, a controller at the bottom level of the hierarchy controls a number of switches referred by the network domain of that controller. A controller at a higher level has multiple subordinate controllers and its network domain is the collection of the network domains of the subordinate controllers. Each controller can only handle requests with both the source and destination inside its domain. Therefore, controllers at higher levels can handle more types of requests.

To model the dynamic decisions of the control plane, we can divide time into slots $t = 1, 2, \ldots$. The status of a controller i at slot t is denoted by $x_i(t) \in \{0, 1\}$ where $x_i(t) = 1$ means controller i is active; otherwise zero. The notation $y_{si}(t) \in \{0, 1\}$ indicates whether switch s is assigned to controller i at slot t or not. Note how these variables extend the controller placement and switch association decisions for many time slots. We also denote with $f_i^k(t)$ the number of CPU cycles that controller i schedules to process the requests of type k at time slot t. The notation $d_{ij}^k(t)$ denotes the number of the requests of type k dispatched to controller j from controller i at slot t. As a result, the queue backlog of the requests of type k at controller i to be processed as per slot t, $Q_i^k(t)$, can be updated by the following equation:

$$Q_i^k(t+1) = \max \left\{ Q_i^k(t) - x_i(t) f_i^k(t)/\rho_k - \sum_j d_{ij}^k(t), 0 \right\} \\ + \sum_s y_{si}(t) R_s^k(t) + \sum_j d_{ji}^k(t),$$

(2.44)

where ρ_k denotes the CPU cycles required to process a request of type k. Note that the first term in the above equation gives the number of unsatisfied requests in the queue at the end of slot t after some requests have been processed at controller i or dispatched to its superordinate controllers. The second and third terms stand for new requests coming from the switches or dispatched from the subordinate controllers.

A desirable property in the above model is to keep the stability of the queues, with the following constraint:

$$\overline{Q_i^k(t)} \leq \infty,$$

(2.45)

where $\overline{A(t)} = \lim_{T \to \infty} \frac{1}{T} \sum_{\tau=0}^{T-1} \mathbb{E}[A(\tau)]$ denotes the time average of a process $A(t)$. Then, subject to the queue stability constraint, the goal of the online optimization method can be to minimize the time-average cost $\overline{\Phi(x(t), y(t), d(t), f(t))}$ due to the running of the controllers, the processing and dispatching of requests. Specifically, the cost at slot t of running the controller and processing requests can be modeled by:

$$\phi_i(t) = x_i(t) \left[\zeta_i^{on} + \zeta_i(t) \sum_k f_i^k(t) \right] + (1 - x_i(t)) \zeta_i^{off},$$

(2.46)

where ζ_i^{on} and ζ_i^{off} denote the cost of controller i to keep its processor active or idle, respectively. $\zeta_i^{(t)}$ denotes the computing cost at controller i at time t per CPU cycle. Similarly, dispatching requests between controllers i and j induces a cost due to the bandwidth consumed of the links connecting the controllers which can be modeled by:

$$\phi_{ij}(t) = \zeta_{ij}(t) \sum_k d_{ij}^k(t),$$

(2.47)

where $\zeta_{ij}(t)$ denotes the cost (e.g., energy consumption) of transmitting the request over the link. For example, the size of a PacketIn message is 256 bits in OpenFlow. The total cost is the sum of the aforementioned costs, $\Phi(t) = \sum_i \phi_i(t) + \sum_{ij} \phi_{ij}(t)$.

We emphasize that the optimal solution to the above problem would require the *a priori* knowledge on request arrivals and computing resources across the network for all time slots. This is not a realistic assumption to make though since in practice traffic often varies in an unpredictable manner. Besides, the decision variables are coupled from slot to slot. Therefore, myopic optimization that greedily minimizes the cost per slot would be strictly suboptimal. The problem can be rather solved using a stochastic subgradient method as described in [83]. Relying on Lyapunov optimization techniques, the asymptotic optimality of such methods can be shown.

For completeness, we note that along with the controller activation, controller-switch association, request processing, and request dispatching to other controllers, there exists one more degree of freedom when designing an adaptive control plane. Specifically, the control plane can *devolve* some of the requests back to the switches so as to mitigate control plane's workloads and fully utilize switches' processing power. Of course, not all the types of requests can be devolved to the switches given their weaker CPU power compared to the controllers. Besides, to be able to devolve more types of requests to the switches the latter will need to maintain state information necessary to process the requests that is normally maintained only at the controllers. Some recent works proposed such flexible state distribution methods to enable the devolution of requests to switches, e.g., see [84]. The devolution decisions can be incorporated into the online optimization framework outlined above to enable an even more flexible control plane, utilizing also the capabilities of the switches in the data plane. A first attempt toward this direction was made in [85] where again Lyapunov optimization techniques were used to transform the long-term stochastic optimization problem into a drift-plus-penalty maximization problem. Yet, this work considered only the controller-switch association and devolution decisions, neglecting the controller activation and request processing/dispatching decisions.

We should also emphasize that another benefit of these type of techniques is that they are amenable to distributed implementation since each controller/switch requires only instantly probed information about queue backlogs and communication/computation costs to carry out its own decision making. This is important to ensure the scalability and reliability. Another important observation is that it may be more practical to make some of the above decisions at different time scales. Specifically, we may allow the (de)activation of controllers and controller-switch (re)associations to be carried out at slot $t = mT$ with a T-slot interval, where $T \geq 1$ and $m = 0, 1, 2, \ldots$. By this, we avoid frequent (de)activation and association of controllers, hence preventing frequent disruptions resulting from the update of network knowledge (e.g., flow tables) following the (de)activation and association of the controllers. Still, the request processing/dispatching/devolution decision making can be carried out at per time slot basis.

We finally note that the adaptive control plane problem was defined above for a hierarchical architecture of controllers. This architecture presents more opportunities for (de)activation of controllers as typically the same switch is managed by multiple controllers at different levels, so there is some redundancy which leads to more degrees of freedom in decision making. The problem can also be defined for a flat architecture of controllers in a similar way (without the decisions on request dispatching among controllers) where the same solution methods can be applied.

Table 2.6: State-of-the-art work on wireless controller placement

Ref.	Network Type	Objective	Method
[87]	Wireless	min # controllers s.t. latency	chance constrained stochastic programming
[88]	WiFi	min latency, probability of failure and transparency	k-Median, Simulated Annealing
[89]	Multi-hop Wireless	max throughput	Lyapunov optimization
[90]	5G	min latency, probability of failure and transparency	Constrained k-means clustering
[91]	Drone	min # controllers, path loss of controller-switch connections and # hops of inter-controller connections	pathloss- and capacity-aware greedy

2.1.6   WIRELESS NETWORK APPLICATIONS

While most of the work on controller placement was primarily in the context of wired networks (wide-area and data center networks), a more recent trend is to explore this problem in a wireless network context. Table 2.6 summarizes the state-of-the-art work on wireless controller placement. This thrust of work is motivated by various software-based network architecture solutions which are realized by the SDN and NFV paradigms. Examples of such architectures include OpenRadio, SoftRAN, and SoftAir, to name a few (see [86] for an extensive survey).

How Is This Different Than Wired Networks?

The SDN controllers in these architectures can be placed in servers which may be located in the mobile core network or even inside the wireless infrastructure network by leveraging edge-cloud technologies. For generality, we assume that controllers may have the flexibility to use wireless links, backhaul or fronthaul, to communicate with each other and with the data plane switches, as shown in Figure 2.8. *The shared wireless medium behaves differently from the wired one since it has to handle radio interference, noise, fading signals, and other RF phenomena.* These factors are not considered by the work on wired controller placement where typically decisions are made based on a latency metric modeled as the Euclidean distance of the multi-hop path between the switch and its assigned controller. However, latency in wireless networks requires a different model which in turn requires a good understanding of the wireless propagation effects. Besides, metrics other than latency may be more relevant and important in the wireless network such as link failure probability that is negligible in wired networks.

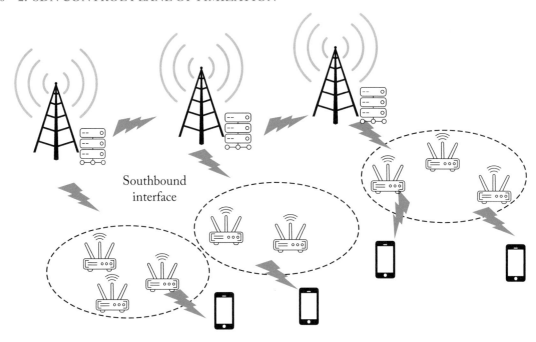

Figure 2.8: Illustration of wireless SDN. Wireless Access Points (APs) in the data plane are logically divided into domains A, B, and C and are associated with controllers A, B, and C, respectively. Communication between the controllers and APs takes place through wireless links.

Modeling the Wireless Channel Delay

The work in [87] made a first attempt to model the controller placement problem taking into account the aforementioned wireless factors. For each wireless link between a controller c and a switch s, a probability p_{sc} of successful transmission over this link is introduced. This probability can capture both path loss and shadowing factors if defined as:

$$p_{sc} = Q \left(\frac{P_{\min} - (P_t + 10 \log_{10} k - 10\gamma \log_{10}(d/d_0))}{\sigma_{\psi_{dB}}} \right), \qquad (2.48)$$

where $Q(.)$ is the Q-function, $P_{\min}$ is the minimum received power below which an outage occurs (in dBm), P_t is the transmitted power (in dBm), k is the unitless constant that depends on the antenna characteristics and the average channel attenuation, γ is the path-loss exponent, d is the distance between controller c and switch s, d_0 is a reference distance for the antenna far field, and $\sigma_{\psi_{dB}}$ is the standard deviation of the Gaussian distributed variable ψ_{dB}.

In the case that a message sent by switch s is not successfully delivered to the controller c, s will retransmit its message to c. The number of retransmissions until the message is successfully received, denoted by η_{sc}, is geometrically distributed as follows:

$$\Pr\{\eta_{sc} = i\} = p_{sc}(1 - p_{sc})^{i-1}, \ i \in \{1, 2, 3, \ldots\}. \tag{2.49}$$

Given the above equation, we can define the propagation delay between controller c and switch s as: $D^{prop} = 2\eta_{sc}t_{sc}$ where the factor 2 is because we compute the roundtrip time and the term t_{sc} is a constant. Note that this is a different propagation delay model than the one in wired networks where the probability of failed transmission is very low.

The authors further enrich the delay model by defining the network access delay. Specifically, they assume that all switches assigned to the same controller will communicate with it over the same frequency channel but during different TDMA slots. Therefore, a switch will suffer a certain delay before getting access to the channel. The expected network access delay for a switch associated with a controller v is given by:

$$D^{access} = \sum_{i=0}^{\sum_{s \in V} y_{sc} - 1} (i T \Pr\{D = iT\}), \tag{2.50}$$

where T denotes the duration of a time slot. The delay increases with the number of switches associated with controller v which is captured by the association variables $\{y_{sc}\}$. Specifically, the first switch will experience delay T, the second delay $2T$, and so on. Each switch will have a probability $\Pr\{D = iT\}$ to be the ith one to transmit.

Given the above models of the number of re-transmissions and access delay, the authors formulate the controller placement problem with the objective of minimizing the number of placed controllers subject to a delay constraint. Since the number of re-transmissions is a random variable, so is the propagation delay. Therefore, the delay constraint is on the probability that the delay will be above a threshold value. Such types of problems are typically referred to as chance-constrained stochastic programs. To address the problem, the authors propose an equivalent deterministic formulation which allows solving it using standard integer programming solvers.

The Aloha Case

The work in [88] made another attempt to model the wireless controller placement problem. For a WiFi network like the one depicted in Figure 2.8, this work considered as objective a function which is a weighted sum of three objectives: (i) latency, (ii) link failure probability, and (iii) transparency. Latency was modeled by assuming *a Rayleigh fading system with slotted ALOHA for channel access.* The link failure probability is the probability that the Signal to Noise Ratio (SNR) of a link between an AP and its assigned controller would go below a threshold. The transparency is a new metric defined as the marginal latency on the data plane caused by the co-channel interference that is added by the control plane. The problem is formulated as a K-Median problem which is NP-Hard. To tackle large problem instances, the authors also propose a heuristic that is based on the simulated annealing algorithm. At a high level, algorithms of this type are probabilistic and generate worse solutions with some probability. The algorithm uses a control parameter known as temperature such that the probability of accepting worse solutions decreases with the temperature. Hence, it allows the algorithm to explore the search space at higher temperatures and helps in the convergence at lower temperatures. Furthermore, the probability of accepting worse solutions decreases with the difference between objectives of current and new solutions. The input to the algorithm is the network graph and the annealing schedule. The algorithm returns the best solution encountered during the traversal.

Connection to the Packet Scheduling Problem
The work in [89] studied the problem of placing a controller to schedule the transmission of data packets in a wireless multi-hop network. The controller collects state information from the rest of the network about the conditions of the links and the lengths of the queues where packets are accumulated before transmission from node to node. Although SDN is not explicitly mentioned, the controller can be regarded as an SDN controller and the rest of the nodes as data plane switches, with scheduling being a control application programmed inside the SDN controller. The key observation is that the delay between a controller node and the data plane nodes increases with each hop between them, and this delay can lead to outdated state information collected by the controller which in turn can impact the throughput performance of the scheduling algorithm. Therefore, the placement of the controller affects network performance.

    With the above observation in mind, the authors propose a new formulation of the controller placement problem. Intuitively, the controller achieves high throughput for data transmissions over links near it but lower throughput for links that are further away due to the increased delay in the state information. This motivates the use of a dynamic controller placement method, in which the controller is allowed to be moved to a region of the network with high queue lengths to increase throughput to this region and provide network stability (bounded queue lengths). The authors in [89] propose such a dynamic method for the case that the channel state evolves over time according to a Markov chain as the one shown in Figure 2.9a. Here, each channel has two possible states, ON and OFF, and the transition probabilities from one state to another are assumed to be fixed and known. The authors then argue that the scheduling decisions can be represented by a matching in the network which is a subset of links which do not interfere with one another and, therefore, can be scheduled simultaneously. Figure 2.9b illustrates an example wireless system where only one of the nodes can transmit at each time due to interference constraints. Therefore, the matching here consists of only one link. Let $\mathcal{M}$ be the set of all possible matchings in the network. Under a throughput maximization objective, a controller placed at node u should schedule the matching $M \in \mathcal{M}$ that maximizes the expected sum-rate throughput with respect to the state collection delays at each node:

$$\max_{M \in \mathcal{M}} \sum_{l \in M} \mathbb{E}[S_l(t) | S_l(t - d_u(l))], \qquad (2.51)$$

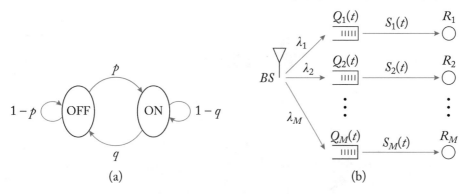

Figure 2.9: (a) Markov Chain describing the channel state evolution of each independent channel in [89]. (b) An example wireless system. Packets are accumulated in queues $Q(t)$ before transmitted to receiver nodes on the right. Due to interference only one wireless link can be activated at each time, provided that its current state $S(t)$ is ON.

where $S_l(t)$ denotes the channel state of link l at time slot t. The term $d_u(l)$ denotes the delay of collecting the state of link l by the controller at node u. The above expression captures how the delayed view of the controller on the state of the links ($S_l(t - d_u(l))$ values) affects the expected throughput. The optimal controller placement is simply to select the node $u \in V$ that yields the highest expected throughput based on the above expression.

5G Networks with Frequent Handoffs

The work in [90] studied the controller placement problem in the context of 5th generation cellular networks (5G). Compared to the 4G LTE standard, 5G brings new design paradigms related to the disaggregation of the base station, which is referred by 3GPP as the Next Generation Node Base (gNB). Specifically, the 3GPP has proposed different splits of the gNB protocol stack with the lower layers in Distributed Units (DUs) on poles and towers and the higher layers in Centralized Units (CUs) which can be hosted in edge clouds. The authors propose a two-layer architecture of SDN controllers to collect and process network data, as shown in Figure 2.10. At the bottom layer, each Radio Access Network (RAN) controller is associated with a cluster of base stations, also referred by Radio Units (RUs). Each RAN controller manages user-level connectivity by coordinating handover decisions and performing load balancing or other QoS policies. The Cloud Network Controller resides at a remote cloud facility and orchestrates RAN controllers.

We should emphasize here the different timescales on which decisions are made in different layers. The DUs schedule over the air transmissions on a sub-ms basis. The RAN controllers may decide upon handover decisions on a timescale of tens of msec. Finally, the Cloud Network Controller can operate on multiple-second intervals to change the association between

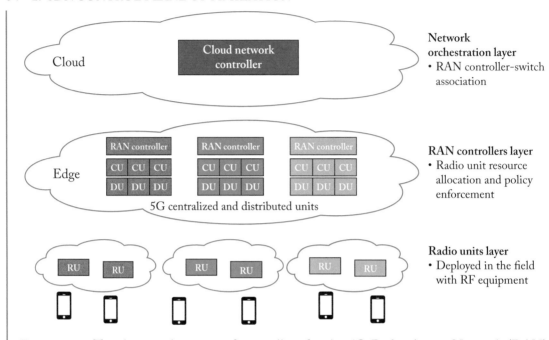

Figure 2.10: Two-layer architecture of controllers for the 5G Radio Access Network (RAN). Each RAN controller is associated with a cluster of Radio Units (RUs) while the Cloud Network Controller orchestrates the RAN controllers [90].

RAN controllers and base stations. The association problem is part of the controller placement and has been extensively studied in the context of wired networks as discussed in the previous sections. However, 5G networks have characteristics that introduce new dimensions which are mainly related to the mobility of the users of such networks. Since the RAN controllers are used to manage user sessions and handover events of mobile users, they will need to maintain a consistent state for each user associated with each base station. When a mobile user handovers between two base stations that are managed by different RAN controllers, these two RAN controllers will have to exchange state information about that user to handle the handover. When deciding the association between RAN controllers and base stations, one consideration is to minimize the number of times the users handover between two base stations associated with different controllers.

The authors propose a data-driven approach to address this. The proposed scheme is run at the Cloud Network Controller which collects time-series of handovers for all base stations and builds a transition probability matrix where the transition probability between two base stations is simply the sample mean of the respective handovers. These probabilities are used as weights in the network graph, and a clustering algorithm is used to identify the associations between RAN controllers and base stations. This algorithm clusters the graph so that intra-cluster handovers are more frequent than inter-cluster ones. To achieve this, a variant of standard spectral clustering techniques for graphs is used, which relies on a constrained version of the K-means algorithm. The algorithm takes as input the weight matrix of the graph $W(G)$, computes the diagonal matrix D so that $D_{i,i} = \sum_j W(G)_{i,j}$, and then computes the normalized graph Laplacian as $L = I - D^{-1}W(G)$ and extracts the eigenvectors associated to the smallest eigenvalues as columns. Finally, it applies the constrained k-means on the rows of the matrix formed by the eigenvectors. The output is a number of clusters, one for each RAN controller, which determine the controller-base station association decisions.

Drone Networks

The work in [91] considered a drone network deployed to provide communication and connectivity at a remote area where ground cellular infrastructure is not available. Examples of such cases include missions targeting natural disasters and remote scientific missions where some form of communication is needed. In such scenarios, the drones have limited degree of mobility, yet their locations may change at some points in time to accommodate new mission requirements. The drones are SDN-enabled meaning that they run SDN-compatible software such as OpenFlow. Some of the drones may also run control plane software and act as controllers. During the operation of the network, the formation of drones may change and thus the controller-drone association may have to change. In case of major changes in the network, it may make sense to update the controller locations as well. Therefore, a dynamic controller placement may be a better fit in this type of network. An example drone network is illustrated in Figure 2.11.

To model the dynamic controller placement problem, the work in [91] presented a general model where controllers can be deployed anywhere in the 2D plane at a suitable altitude, as opposed to most of the work discussed in the previous sections where controllers are assumed to be located within any of the nodes in the network topology. A joint controller placement and switch association problem is formulated as an ILP in which the objective is to minimize a weighted sum of the number of controllers, the path loss of the connections between controllers and switches, and the number of hops of the inter-controller connections. An interesting point is the modeling of the link path loss metric. Specifically, the authors use the Fris-based free space propagation model, where path loss, in dB, over transmission distance d is defined as:

$$Pl(d) = Pl(d_0) + \eta \log_{10}\left(\frac{d}{d_0}\right), \tag{2.52}$$

where d_0 is the reference distance (1 m) and η is the pathloss exponent ($\eta = 3$), and $Pl(d_0)$ is the Fris transmission equation defined as:

$$Pl(d_0) = \frac{P_t G_t G_r \lambda^2}{16\pi^2 d_0^2 L}, \tag{2.53}$$

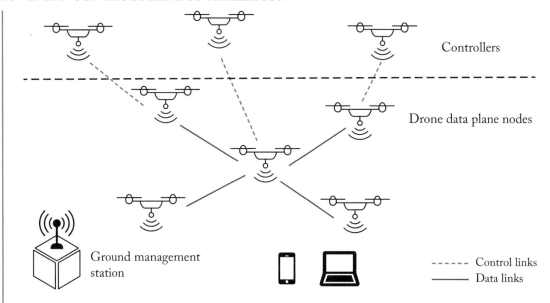

Figure 2.11: Illustration of controller placement in SDN-enabled drone network. A dynamic strategy is needed to handle network changes due to the mobility of the drones [91].

where P_t is the transmission power, G_t and G_r are the transmission and reception gains (1 dBm), λ is the wave length, and L is the system loss (equals 1).

The problem is formulated subject to constraints on the computation capacity of controllers and path loss values of the controller-switch connections. Since the problem is an ILP, it can be solved for small and medium sizes using standard ILP solvers like CPLEX. The extra challenge here is that the solution should be updated over time to adapt to network changes (drone movements). Repeatedly solving the ILP problem and updating the association and placement decisions accordingly would require an arbitrarily high number of re-associations and movements of controllers which is undesirable. Instead, the authors propose a heuristic algorithm that makes incremental changes only to the association decisions, and solves the ILP problem to update the controller locations less often, only when major network changes have happened.

The heuristic algorithm is greedy in nature and updates the association decisions based on the pathloss values and residual capacities of the controllers. This algorithm operates in two phases. In the first phase, it updates the associations between drones and controllers without changing the controller locations. In the second phase, it updates the controller locations if needed (if some drones cannot find a controller with good enough path loss value and sufficient available capacity). In detail, the first phase attempts to re-associate drones that have moved to new locations to existing controllers. It traverses these drones one by one to check if their currently associated controller still provides a good enough path loss value. If not, it creates a list of controllers and ranks them with respect to their path loss value and keeps only the controllers that satisfy the path loss constraint. The first controller is then picked that can also satisfy the capacity constraint. If no such controller can be found (the list is empty), then the second phase of the algorithm is executed. In the second phase, the ILP is solved again. This time though the objective changes and instead of minimizing the number of controllers term it minimizes the distances of the new controllers we place to the existing controllers. This way we ensure the new placement will not be much different than the existing one, and thus the energy and time overheads of moving the controllers to the new locations will be kept to the minimum.

This is the last work on controller placement that we discuss in this chapter. We remark that wireless applications of SDN is a relatively recent topic and the application to drone networks is a rather futuristic case. We will continue the discussion of such futuristic applications of SDN in mobile ad hoc networks in Chapter 4.

2.2   CONTROL TRAFFIC MANAGEMENT

While most of the SDN research activity has been focused on the controller placement problem discussed above, there is also an increasing interest in the management of the traffic the controllers inject into the network during their operation. This traffic, from now on referred to as control traffic, contains two main components: (i) the traffic between the controllers and their associated switches for state collection and control operation and (ii) the traffic exchanged between the controllers (when multiple controllers have been placed in the network) for state synchronization (i.e., for synchronizing the local views of the controllers on the states of switches). Since the control traffic consumes network resources and may reveal information about the network applications, the management of this traffic should be optimized accordingly. In the following, we present example evaluation results of the control traffic. The results are based on two real commercial SDN controller implementations. Following that, we present methods for managing the two types of control traffic to address the aforementioned challenges.

2.2.1   EVALUATION OF CONTROL TRAFFIC

To get an understanding of the magnitude of the control traffic, we present numerical evaluation results of two real SDN controller implementations—ONOS and OpenDaylight. The two implementations employ different strategies for state synchronization among controllers, and therefore the amounts of control traffic they inject into the network are different. The main difference is that OpenDaylight adopts a leader-based synchronization approach where one of the controllers, the leader, has a designated role in the synchronization process. In contrast, ONOS adopts a more symmetric (flat) state synchronization approach.

In more detail, OpenDaylight adopts the Raft consensus algorithm to achieve state synchronization among the controllers. One of the controllers is elected to be the leader and has the responsibility to update the state and replicate it to the other controllers (followers). So, all the state updates in the whole network are handled by the leader, resulting in heavy load to that controller compared to the rest. On the other hand, ONOS adopts multiple protocols for synchronization. The first one is again RAFT. ONOS uses this protocol to synchronize the controller-switch associations so that all controllers agree on them. The second is the anti-entropy gossip protocol. This is used to synchronize information about network topology and flow tables among controllers. Differently from RAFT, the synchronization process in this protocol is leader-less. Periodically, a random controller is chosen to send the state of its domain (i.e., switches associated with that controller) to another random controller.

A practical way to evaluate the control traffic of OpenDaylight and ONOS is by using an emulation platform connected with the software of these two controller implementations. Mininet [92] is such a platform that has been extensively used by SDN researchers. Note that other popular platforms like NS3 could also be used but they are not that easy (although it is possible) to connect with SDN software. With Mininet, we can easily evaluate networks with hundreds of nodes and several controllers. This evaluation method was followed by the work in [59, 60]. The authors in this work emulated the operation of an edge network (e.g., a wireless access network) with ring topology and evenly associated nodes with the placed controllers, as illustrated in Figure 2.12a. All the controllers ran a simple reactive forwarding application.

Emulations were performed separately for ONOS and OpenDaylight. Figure 2.12b shows the results for ONOS. In order to show the impact of the scale of the network, the figure shows both types of control traffic (controller-node and inter-controller) for different numbers of nodes in the virtual network. The figure verifies that both the types of traffic grow linearly when the network scales up. What is more, other factors that have impact are considered. For controller-node traffic, a large amount of one-hop *iperf* flows are created randomly, with a fixed rate (0.1 flows per second for each node). For inter-controller traffic, different numbers of controllers (cluster sizes) are placed. According to this figure, in all of these situations, the two types of overheads are at the same order of magnitude (up to a few Mbps each). This fact means that both of them are important in order to make the control plane design choices. Figure 2.12c elaborates on the different types of messages exchanged between the controllers in ONOS. Specifically, the controllers exchange periodic heartbeat messages of fixed number and size. Therefore, the respective traffic overheads are independent of the number of data plane nodes. Besides, the RAFT protocol (that synchronizes controller-node associations) and the anti-entropy gossip protocol (that synchronizes topology and flow tables) are used. These overheads steadily increase with the size of the network.

For the emulations with OpenDaylight, the OpenFlow protocol was adopted for the controller-node communications as with ONOS, leading to similar results. However, the inter-controller traffic pattern is quite different. As shown in Figure 2.12d, although the overhead also grows linearly with the number of data plane nodes, the overhead is much larger compared with the same setting in ONOS case. Importantly, the overhead is no longer evenly distributed among each pair of controllers. In the three-controller case, we can identify one controller as

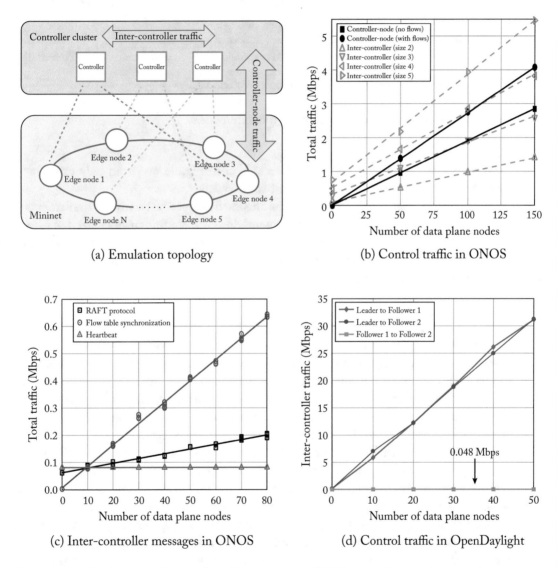

(a) Emulation topology

(b) Control traffic in ONOS

(c) Inter-controller messages in ONOS

(d) Control traffic in OpenDaylight

Figure 2.12: Evaluation of control traffic for two real SDN controller implementations.

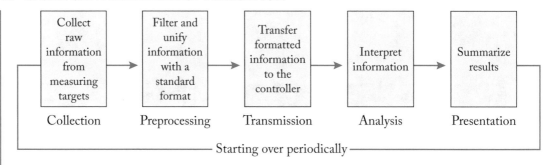

Figure 2.13: Operation phases in network measurement.

a leader, and the remaining two as followers. Non-negligible overhead only exists between a leader and a follower, but not between the two followers.

We remark that the above evaluations are based on specific controller implementations in late 2018. As new versions of these implementations are continuously realized, e.g., adopting new protocols for signalling and synchronization among controllers, the reported values will become outdated. Still, the presented evaluations are important to get an initial understanding of the traffic overheads of the different controllers at the time of writing.

2.2.2   STATE COLLECTION BY THE CONTROLLER

The evaluation results presented above demonstrate that the control traffic volume can be relatively high (of the order of Mbps). This may be significant for some networks where bandwidths to the controller are constrained. This motivated researchers to investigate methods for the management of this traffic. In the following, we discuss such methods particularly targeting the traffic exchanged between the centralized controller and the switches distributed in the data plane network. We refer to this problem as state collection by the controller.

Collecting information about the state of the network by the centralized SDN controller is vital for various network management tasks and applications. For example, the performance of traffic engineering heavily relies on the accuracy and recentness of link load information in the network. Monitoring the value of flow parameters such as bandwidth, packet loss, and latency is also critical for detecting network events in time (e.g., QoS performance degradation) and reacting swiftly to them. Developing mechanisms for transferring such distributed across the network information to the controller is in fact another step in the long history of efforts in the field of network measurement.

Network measurement can be classified into five phases as illustrated in Figure 2.13. First, network devices (switches) run agents that periodically record traffic information in the collection phase. Then is the preprocessing phase where devices distinguish valid from invalid data. Decisions can be sent back to the collection agents to filter out duplicate data from neighborhood network devices. In the transmission phase, valid data from the agents will be sent for analysis to an aggregator server which in the case of SDN is the controller. After being transmitted to the controller, the measurement data will be analyzed (analysis phase) and summarized (presentation phase) so as to pass appropriate inputs to network applications through the northbound interfaces.

Network measurement is an active area of research since the early stages of the Internet. Various methods have been developed to optimize its operation. These methods are divided into two main categories; *active* and *passive*. In active measurement, special probe packets are injected into the network with the aim of measuring the performance of network paths, e.g., round trip times. The granularity of probe packet injection can be adjusted to better meet the needs of applications. However, the probe packets impose significant measurement overhead that may disturb critical traffic flows. On the other hand, passive measurement does not generate any additional traffic. Rather, it captures and analyzes existing traffic at predefined points in the network. This method often depends on packet-sampling mechanisms and employs statistical analysis tools to infer the state of the traffic. The main challenge with this method is that of limiting the measurement inaccuracy by smart sampling strategies since traffic sampling by definition cannot measure all packets of each flow.

The emergence of SDN made network measurements and their use for inference easier. With SDN, the measured network state values are aggregated at the centralized controller through programmable southbound interfaces (e.g., OpenFlow). Therefore, flexible self-tuning mechanisms that monitor flow statistics and dynamically adapt their operation (e.g., monitoring rate) as needed can be programmed with relative ease as SDN applications at the controller. Traditional (non-SDN) measurement methods, both active and passive, can be re-designed with SDN, this time with a more programmable and dynamically re-configurable flavor.

The OpenFlow protocol defines specific methods for the collection of traffic statistics by the switches [96]. The OpenFlow switches maintain traffic statistics in their flow tables such as byte and packet counters for their active flows. In the active method (also referred as "pull-based"), the controller sends Read-State messages through the OpenFlow interface to retrieve the flow statistics from the switches. In the passive method (also referred as "push-based"), the controller relies on the reports from the switches sent through the same interface every time a new flow arrives (at the flow entry installation time) or leaves (at the flow entry timeout) the switch. However, there is no other notification to the controller at any other time during the lifetime of the flow.

An implementation in Mininet of the active and passive measurement methods of Open-Flow was provided in [97]. This work showed that for a load-balancing application running on an SDN controller a passive method that keeps track of the flow rules inserted into and deleted from the switches to estimate link utilizations can actually perform better than an active method where the controller asynchronously communicates with the switches to request the link utilization state information. However, the result of this comparison may be different in other networks with different traffic dynamics and for different applications and so there is no clear winner between the two methods.

Based on the above discussion it becomes clear that deciding the right method to collect network state measurements by the switches and transmit to the SDN controller is not straight-forward. Various methods, active or passive, have been proposed in the SDN literature since 2008, and comprehensive surveys can be found in [98] and [99]. Most of these works proposed active methods where the controller pulls flow statistics from the switches. Two basic questions in this context are the following.

1. How often and when exactly should the controller query the switches for pulling their measurements?

2. When the controller queries a switch, which specific measurements to pull from it if not all the available ones?

To answer the first question, OpenNetMon [100] proposed a method that queries switches at an adaptive rate that increases when flow rates differ between samples and decreases when flows stabilize to minimize the number of queries. The adaptive rate reduces network and switch CPU overhead while optimizing measurement accuracy. To answer the second question, two representative methods were proposed in [101] and [103]. The two works targeted differ-ent problems. The first work aimed at estimating the traffic matrix in a network by collecting statistics for all flows (flow counters) from a subset of the switches. The second work aimed at estimating the link loads by collecting part of the link load utilization information in the network. Both these quantities (traffic matrix and link loads) are important to estimate and is input to traffic engineering and other network control algorithms that we will discuss in detail in the next chapter. In the following, we elaborate on the methods and models used by these two works.

Traffic Matrix Estimation

The problem of collecting flow statistics to estimate the traffic matrix (TM) was modeled as a *graph covering problem* in [101]. The network topology is represented by a graph $\mathcal{G} = (\mathcal{V}, \mathcal{E})$ with $\mathcal{V}$ switches and $\mathcal{E}$ links among them. The TM represents the volume of traffic for all flows that exist in the network. It is reasonable to assume that the set of flows $\mathcal{F}$ is known to the controller since this information is obtainable from the switches. Since the routing path for each flow is determined by the controller, the latter also knows the subset of flows $\mathcal{F}_v$ that pass through each switch v.

A novel aspect of this work is that the controller can send a Read-State command to a switch v that contains a *wildcard rule r* (in a set $\mathcal{R}$ of wildcard rules). This wildcard rule matches a subset of the flows of that switch denoted by $\mathcal{F}_v^r$. For example, a simple way for setting wildcards is to specify the destination, and match all the sources. The switch v will assemble the statistics of the flow entries matching that wildcard rule ($\mathcal{F}_v^r$ set) into a reply packet and send it to the controller. Therefore, this mechanism pulls information from a switch for not necessarily all flows but for only a subset of them. This is useful because for a flow traversing multiple switches it would be redundant to measure it multiple times at different switches.

Another interesting point in this work is the cost model it uses. Specifically, the cost of a wildcard rule r applied on switch v is denoted by $c(\mathcal{F}_v^r)$. This function is defined as $c_1|\mathcal{F}_v^r| + c_2$, where c_1 and c_2 are constants. The function captures the bandwidth overhead of the request and reply packets. The request packet has length 114 bytes while for the reply packet it depends on the number of matched flows. Specifically, the length of the reply packet starts with 74 bytes and increases by another 96 bytes for each matched flow. Therefore, it is defined as $c(\mathcal{F}_v^r) = 96|\mathcal{F}_v^r| + 188$.

Given the above model [101] formulated the problem of selecting the switches and the wildcard rules to send request messages from the controller:

$$\min \lambda \tag{2.54}$$

$$s.t. \sum_{v \in \mathcal{V}, r \in \mathcal{R} \, : \, f \in \mathcal{F}_v^r} z_v^r \geq 1, \; \forall f \in \mathcal{F} \tag{2.55}$$

$$\sum_{r \in \mathcal{R}} z_v^r c(\mathcal{F}_v^r) \leq \lambda, \; \forall v \in \mathcal{V} \tag{2.56}$$

$$z_v^r \in \{0, 1\}, \; \forall v \in \mathcal{V}, r \in \mathcal{R}, \tag{2.57}$$

where the binary variable z_v^r denotes whether the controller will send a request with wildcard r to switch v or not. The first set of inequalities denotes that every flow will be covere, i.e., statistics for it are collected from at least one switch. The second set of inequalities means that the total cost on each switch should not exceed λ. The objective is to minimize the maximum cost on all the switches (or to achieve the cost balancing on switches), that is, $\min \lambda$.

The above problem contains integer variables and can be shown to be NP-Hard. The work in [101] proposed a randomized rounding based algorithm to approximate the optimal solution. The algorithm begins with solving the linear relaxation of the problem where z_v^r can take any fractional value in [0, 1]. Next, iteratively it randomly picks an uncovered flow and covers it by picking greedily the switch corresponding to the largest value in the fractional solution.

We note at this point that although the approach of covering flows by wildcard rules (corresponding to flow statistics to be collected by the switches) is interesting, this approach has one limitation. It implicitly assumes that all the switches on the routing path of a flow are the same for pulling flow statistics. However, as argued in [102], covering a flow by a switch closer to its destination can actually lead to more accurate estimation of its traffic volume due to the packet loss that the flow may suffer at the intermediate switches. This packet loss cannot be measured by switches in the beginning of the routing path of the flow close to its source node. How to extend the above formulation to capture such practical considerations is an interesting open question.

Link Load Estimation

The work in [103] considered a different problem where the goal is to estimate the loads of the links in a network, rather than the traffic matrix. The main idea this time is that the controller does not need to request the load information from all links but from only a subset of them. The rest link of the loads will be estimated from the measured link loads. The argument why this would work is that in many networks the link loads show spatial correlations. So, the load information of all links can be well represented by a few important links. This way, the overall bandwidth overhead and latency of state collection by the controller can be reduced.

A method that can be used for estimating the loads of all links by only monitoring a subset of them is *compressive sensing*. This method takes as input measurements on network-wide link load over long timescales (e.g., days). These measurements can be represented by a matrix $M \in \mathbb{R}^{t \times m}$ where t refers to the number of measurements or snapshots and m is the number of links. Then, we can perform the Singular Value Decomposition (SVD) on the matrix M to check the property of spatial correlation. The SVD method is actually a decomposition of M in the form:

$$M = U \Delta V^T, \tag{2.58}$$

where U and V are called the left and right singular matrices so that $U^T U = V^T V = I$ while Δ is a diagonal matrix whose diagonal values are known as the singular values. An important observation is that often among the m singular values only q of them can have positive value. This means that we can exactly represent the load information of every link by a linear combination of the respective q links without any loss. Further, in many cases, the singular values are sparsely distributed. In other words, there are only a few large singular values, say $k << q$ values, and the rest of them are very small. Thus, every column of matrix M can be approximately represented by only k basic columns, which in turn means that the load of the respective k links can approximately recover the entire load matrix.

To estimate the loads of non-measured links, a linear regression model was proposed in [103]. Evaluations on tens of topologies from the Internet Topology Zoo online repository showed that the estimation error of this simple model is less than 10% in more than 80% of the topologies. The problem can also be solved by more sophisticated statistical models such as nonlinear SVM to reduce further the estimation error.

2.2.3   SYNCHRONIZATION AMONG THE CONTROLLERS

In this section, we focus on the management of the other part of the control traffic, the traffic exchanged among the controllers for state synchronization. We refer to the optimization of this traffic as the controller synchronization problem. We emphasize though that the synchronization among controllers has received much less attention than the controller placement problem. The area is still open for research and innovation.

Before introducing the controller synchronization problem, we remind that when multiple controllers are placed in a network each of them is usually responsible for only a part of the network (a subset of the switches), commonly referred to as the controller's *domain*. The messages disseminated by a controller to the other controllers convey its view of the state of its domain (e.g., available links and installed flows). The composition of these messages allow the controllers to synchronize and agree on the state of the entire network. Most of the work has looked at synchronization among controllers with a focus on rapid and practical protocol implementation, rather than modeling and optimization. Therefore, theoretical work is limited. For example, OpenDaylight and ONOS, two state-of-the-art controller platforms, adopt the RAFT and anti-entropy gossip protocols for disseminating coordination messages among controllers [93]. These two protocols are engineered to achieve consistency among controllers. However, their operation is based on empirical rules and not explicitly optimized.

To better understand controller synchronization, we note that the RAFT and anti-entropy gossip protocols are representative examples of two broad protocol categories. The first category contains the *strongly consistent* protocols (such as RAFT) which strive to maintain all the controllers synchronized at all times. This is ensured by disseminating messages each time a network change (e.g., a node or link failure) happens followed by a consensus procedure. The second category contains the *eventually consistent* protocols (such as anti-entropy gossip) which omit the consensus procedure, and yet converge to a common state in a timely manner usually through periodic message dissemination.

Despite its benefits, strong consistency is difficult to ensure in practice as it is challenged by the unreliable nature of network communications. In addition, this approach generates significant *overheads* for message dissemination among controllers (of the order of Mbps or more as we showed in the previous section) which can be prohibitively large for resource-constrained networks. On the other hand, eventual consistency, where controllers are permitted to temporarily have inconsistent views of each other's state, better suits the needs of the above networks, and thus can be used to extend the applicability of distributed controller solutions. However, inconsistent views of controller states can harm the *performance of network applications*.

One might attempt to model the controller synchronization problem separately for each of the strong and eventual consistency models. This has been attempted so far, for the eventual consistency model, only in a recent series of papers [108–110]. This work was motivated primarily by visionary application of SDN to tactical ad hoc networks, supported by the U.S. and UK Department of Defense (DoD) agencies through the DAIS ITA program [204], where the communication resources for the synchronization among controllers is limited.

The main question to ask in the eventually consistency model is *how often* (at what period or rate) to synchronize each pair of controllers. One might expect that the straightforward policy where all controller pairs synchronize at an equal rate would work well. However, another may argue that the synchronization rate should be higher for domains that are more dynamic (with many changes in topology and flow configurations) as this is needed more in order to preserve the overall consistency among controllers in a network. The issue is further complicated by the requirements of the network applications. The works in [104] and [105] showed that certain network applications, like load-balancers, can work around temporary inconsistency and still deliver acceptable (although degraded) performance. In these cases, some additional effort needs to be made to ensure that conflicts such as forwarding loops, black holes, and reachability violation are avoided [106]. Therefore, synchronization policies that completely neglect the *specific applications of interest* in the network as well as the impact of synchronization rate on their performance may be inefficient.

Synchronization Rate vs. Network Performance

To understand the impact of synchronization rate on network performance, we describe below a brief emulation study originally performed by [108]. This will motivate the optimization methodology as we explain in the sequel. Among the set of commercial controllers that are available online *RYU* is chosen which is open-source and allows to program from scratch new protocols for the synchronization among controllers. Specifically, the authors implement a simple eventually-consistent protocol which periodically disseminates synchronization messages between each controller pair. The code is parameterized to allow for any synchronization period. The disseminated messages convey the local views of the controllers about the topology and installed flow tables. This information is made available to the controllers by the OpenFlow protocol.

The authors first test the performance of a shortest path routing application. With this application, packets are routed to their destination following the path of minimum hop count, calculated by Dijkstra's algorithm. They generate a random network of 16 nodes and 3 controllers, depicted in Figure 2.14a, where links fail or recover randomly every one second with probability 0.05, and nodes with the same color are managed by the same controller. They further generate data packets with random source-destination nodes. Unless the controllers synchronize at the time of packet generation, the packet is at risk of following a failed routing path.

The performance of the routing application is determined by the number of packets that are successfully routed (without attempting to traverse any failed link) to their destinations. The authors emulate the performance for five different scenarios where all the controller pairs synchronize at the same rate equal to (i) 0.5, (ii) 0.25, (iii) 0.125, (iv) 0.063, and (v) 0.031 (messages per second). This translates to a single message disseminated every 2, 4, 8, 16, or 32 seconds. For each scenario, emulations are run for multiple times and the results are depicted in Figure 2.14b. Despite a large extent of randomness, we observe that the average performance (calculated over 20 minutes) increases with the synchronization rate and saturates eventually showing that a *diminishing return rule* applies.

Additional emulations are performed to test the performance of a load balancing application. A similar setup with the work in [104] is considered, depicted in Figure 2.14c. That is, a network is generated with two controllers. Each controller manages two nodes, a switch and a server. The switches generate flows uniformly at random. The flows can be routed and queued to any of the two servers. Each controller is aware of the load of the server it manages. It also

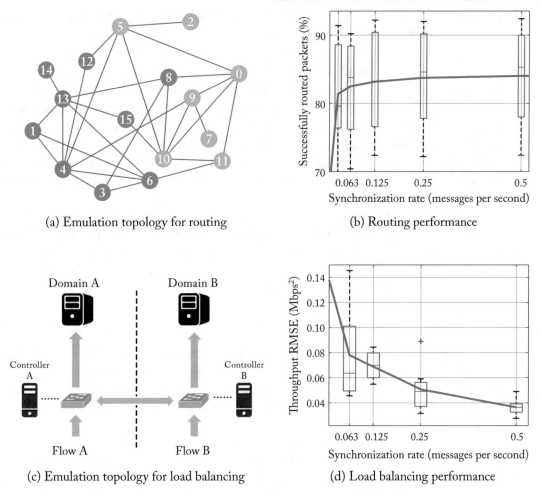

(a) Emulation topology for routing

(b) Routing performance

(c) Emulation topology for load balancing

(d) Load balancing performance

Figure 2.14: Emulation topology and impact of synchronization rate on the performance (box plots and average values) of (a)(b) shortest path routing and (c)(d) load balancing applications.

receives periodic synchronization messages about the load of the other server by the other controller. Each time a new flow is generated, the responsible controller routes it to the server with the currently observed lowest load. However, this may not be the least loaded server in reality, since the controllers are not synchronized at all times.

The emulation results are depicted in Figure 2.14d. The metric depicted is the root-mean-square deviation (RMSE) of two servers' throughputs. The better the two server loads balance, the lower the value of this metric becomes. Therefore, this metric captures the performance of a load balancing application. Coinciding with the routing application, we observe that the performance improves with the synchronization rate but gradually saturates showing that a diminishing return rule applies.

Learning the Optimal Synchronization Rates

The emulation results presented above have shown the impact of synchronization rates on the performance of network applications and provided some insights we can use to optimize these rates. Building upon these results, the work in [108] formally defined the problem of deciding the synchronization rates that maximize the performance of network applications. While the performance function is expected to have a curve shape similar to those reported in the previous figures, we cannot express it in closed-form. Therefore, the objective function is *unknown*, complicating the problem.

To overcome the unknown objective challenge, the authors of [108] proposed to leverage methods from *learning theory*. Such methods typically *train* the system by trying-out a sequence of solutions (synchronization rates) over some training period $\mathcal{T} = \{1, 2, \ldots, T\}$ of T time slots, until they can infer a "sufficiently good" solution. To describe such training process, the vector of synchronization rate variables is defined as:

$$r = (r_{ij}^t \in \{0, 1, \ldots, R\} \; : \; \forall i, j \in \mathcal{C}, j \neq i, t \in \mathcal{T}), \tag{2.59}$$

where r_{ij}^t indicates the synchronization rate between controllers i and j tried-out in time slot t, and R stands for the maximum possible rate. $\mathcal{C}$ is the set of controllers where $C = |\mathcal{C}|$. Note here that r is the fourth basic design choice we can make to optimize the SDN control plane together with the controller placement (x), switch association (y), and statistic collection (z), we discussed in the previous sections.

Given the synchronization rate vector r^t tried-out in a slot t, the application performance will be $\Psi_t(r^t)$. Here, $\Psi_t(.)$ is an unknown function that governs the application performance in slot t. While the overall function is unknown, the single value $\Psi_t(r^t)$ can be *observed* by the system *after* the synchronization rate decision r^t is made, in the end of the slot. For a shortest path routing application, for example, this is possible by measuring the number of data packets that reached their destination in time. Such information can be estimated by the controllers using for example information from Transport Control Protocol (TCP) acknowledgement packets. The information can be then passed to the system (e.g., one of the controllers) which can simply aggregate and sum the respective values.

We emphasize that the function $\Psi_t(.)$ is time slot-dependent, meaning that the performance value might change with time even for the same synchronization rate decision. That is, we may try-out the same synchronization rate vector $r^t = r^{t'}$ in two slots t and t' but observe different performance values $\Psi_t(r^t) \neq \Psi_{t'}(r^{t'})$. Such *uncertainty of observations* is due to the stochastic nature of the network. Intuitively, the performance value will be large if the network happens to be stable in a slot but will be much worse in other slots during which many changes happen.

Despite the uncertainty of observations, the learning method should be able to infer by the end of the training period $\mathcal{T}$ a "sufficiently good" synchronization rate decision $\hat{r} = (\hat{r}_{ij} : i, j \in \mathcal{C}, j \neq i)$. This should, ideally, maximize the *average* performance denoted by an (also unknown) function $\hat{\Psi}(.) = \mathbb{E}[\Psi_t(.)]$. Therefore, the objective can be written as:

$$\max_{\hat{r}} \quad \hat{\Psi}(\hat{r}) \tag{2.60}$$

$$s.t. \quad \sum_{i \in \mathcal{C}} \sum_{j \in \mathcal{C}, j \neq i} \hat{r}_{ij} b_{ij} \leq B, \tag{2.61}$$

where inequality (2.61) ensures that the inferred synchronization rate decision will satisfy the resource constraint (budget B). We note that the average performance $\hat{\Psi}(\hat{r})$ can in fact be the aggregate of many (rather than only one) applications. Either way, performance is not the only criterion that determines the efficiency of a learning method. Another important criterion in this context is the *running (or training) time T*, i.e., how many time slots are required for training in order to infer the synchronization rate decision $\hat{r}$.

To handle the uncertainty of an observed performance value $\Psi_t(r^t)$, a learning method would typically try-out the *same* synchronization decision r^t *multiple times*, in different time slots. Then, the empirical mean of the observations will be used to estimate the average performance value $\hat{\Psi}(r^t)$. By repeating the above training process for every possible synchronization decision, an estimate of the entire objective function $\hat{\Psi}(.)$ can be obtained. However, there exists an *exponential number of possible decisions*; $(R + 1)^{C(C-1)}$ decisions in total. Therefore, this approach would require an exponential number of time slots for training, which is clearly not practical.

To overcome the high dimensionality of the synchronization decision space, we could leverage learning methods proposed recently that do not require the estimation of the objective function at every possible decision. For instance, the *ExpGreedy algorithm* proposed in [107] can infer a close-to-optimal decision in *polynomial-time* provided that the objective function follows a diminishing return rule. Still, however, the running time of this algorithm may be too large for the synchronization problem, as shown numerically in [108].

Based on the above, [108] proposed an alternative more-practical learning algorithm for which we can flexibly adjust the running time by setting appropriate values to its input parameters. In a nutshell, the algorithm starts with the all-zero synchronization decision and then gradually constructs the decision to be returned by iteratively increasing by 1 the synchronization rate of a single controller pair. This procedure will end when the B resource constraint is reached, i.e., after B iterations. Each iteration requires multiple time slots for training so as to be confident that the controller pair selected to increase its rate by 1 will improve the average performance more than other controller pairs.

Interestingly, the algorithm outlined above, referred to as Stochastic Greedy in [108], performs very well even compared to a number of state-of-the-art learning algorithms such as the widely used online convex optimization (OCO) and deep reinforcement learning algorithms (DQN). Figure 2.15 demonstrates the performance benefits when these algorithms are applied

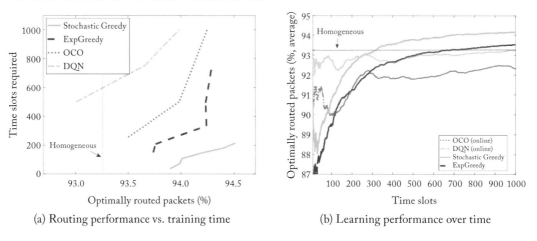

(a) Routing performance vs. training time   (b) Learning performance over time

Figure 2.15: (a) Tradeoff between performance and training time and (b) learning process under the shortest path routing application in [108].

to compute the synchronization rates among three controllers when the application of interest is shortest path routing. Here, OCO generalizes the basic gradient descent method used for standard offline optimization to explore an unknown function. DQN (Deep Q-Network) method deploys a neural network with two fully-connected hidden layers each with 24 neurons and ReLu activation function, and an output layer with a linear activation function to estimate the value of different actions (increase/decrease of synchronization rates) under different states (vector of synchronization rates between each pair of controllers). This type of deep reinforcement learning formulations are attracting increasing interest lately.

Figure 2.15a depicts the detailed tradeoff between performance and training time for all the algorithms. It shows that we can flexibly trade the performance and training time of the Stochastic Greedy algorithm (from 93.9–94.5% optimally routed packets and from 50–210 slots). To achieve the same performance level, ExpGreedy typically takes more than twice the time. As for OCO and DQN, they generally need more training time than greedy algorithms. In order to merely have better performance than the baseline (referred to as "homogeneous") that synchronizes all pairs of controllers at an equal rate, more than 400 time slots are required, limiting their applicability in practical scenarios.

Figure 2.15b examines the performance over time of the different learning algorithms. Specifically, the objective function is tracked for 1000 time slots during which the Stochastic Greedy algorithm is finished first and then the output policy is applied for the rest of the slots. OCO and DQN, are kept in the training stage for all slots. The figure depicts the average value of the objective function during the learning process for the different algorithms. We observe that the online methods (OCO and DQN) are better at first because they start at the homogeneous setting, while the greedy methods (Stochastic Greedy and ExpGreedy) start at the minimum-level setting. For OCO, additional randomness is introduced by the gradient estimation procedure, causing it to deviate from the optimal point frequently which leads to inferior performance. For DQN, the learning process is relatively slow. It can be observed that the average objective function keeps fluctuating in first 500 slots, during which the greedy algorithm has already finished.

While the above constitute an initial attempt to model and solve the synchronization problem in SDN, this area is still quite open for research. First, the problem above is modeled for an eventually consistent system where controllers synchronize periodically, yet at possibly different rates. An interesting question is how the problem can be modeled for a strongly consistent system where synchronization takes place in an event-driven manner. The situation becomes even more complex for controllers like OpenDaylight which adopt both eventually consistent and strongly consistent routines for synchronization based on the RAFT and anti-entropy gossip protocols.

2.2.4   PRIVACY OF THE CONTROLLER

The synchronization problem in the previous section was defined for the case that controllers share their local domain state information (network topology and resources) with each other so that all controllers maintain the same (although sometimes outdated) global state information about all domains. This was motivated by controller implementations like ONOS and Open-Daylight that work based on this principle. However, in certain cases the controllers may be reluctant to expose their domain states to other controllers due to privacy concerns. In this section, we briefly discuss methods for achieving coordination among controllers while preserving their privacy.

Topology aggregation is a popular method for preserving privacy among network domains. The idea is to generate an abstract view of the network topology of each domain where only border switches can be seen by the other domains. The physical topology is thus simplified and transformed to a logical topology where border switches are fully connected through virtual links. Consider for example an SDN network with four domains as in Figure 2.16. On the right side, we depict the aggregate topology seen by the controller of domain #1. This is a combination of virtual links inside domains #2, #3, and #4, as well as of the real links connecting the border switches. Although this topology aggregation method cannot prevent others from getting information about border switches, it is still able to prevent many attacks toward domain-internal switches, as long as their IP addresses are protected.

We remark that we can pass additional network-application dependent information along with the aggregated topology from one controller to another to facilitate their collaboration without sacrificing much of their privacy. For instance, to support a traffic engineering application that provides routing and bandwidth allocation for flows possibly crossing multiple domains, [111] proposed to pass from one controller to the others some delay and bandwidth information about the (real or virtual) links. Let us denote by $< d_l, b_l >$ the delay and bandwidth

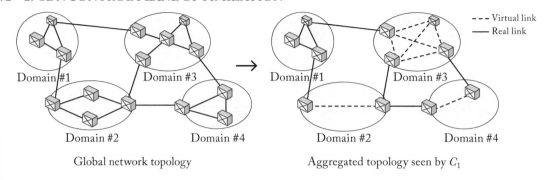

Figure 2.16: Topology aggregation in multi-domain SDN [111]. The actual topologies of the domains (left) vs. the view available to the controller of domain #1 (right).

of a link l. A path can be denoted by a sequence of links $q = l_1, l_2, \ldots, l_{|t|}$. We can compute the bandwidth of this path as $b_q = \min\{b_1, b_2, \ldots, b_{|q|}\}$ and the delay $d_q = \sum_{l \in q} d_l$. For each flow request, the local controller can compute a shortest (minimum delay) path on the aggregated topology that has sufficient bandwidth, e.g., by running the Dijkstra algorithm. The critical point here is that controllers should communicate to each other their bandwidth demands to achieve some level of synchronization and avoid wasting resources when for the same flow more bandwidth is allocated in one domain than in other domains. A mechanism to achieve such synchronization among controllers was presented in [111]. The idea is to compute the minimum available bandwidth for each flow across domains and notify the controllers of other domains to return an amount of bandwidth equal to the deficit. This process can be carried out over several rounds to satisfy the demands of flows with different priorities so as to ensure that each flow receives a fair share of bandwidth. The interested reader is referred to [111] for more details.

CHAPTER 3

SDN Data Plane Optimization

The previous chapter focused on design choices for the SDN control plane such as the placement of controllers and the management of control traffic. Optimizing the data plane is another important issue and hence has been of much research interest. SDN applications running on the control plane require the implementation of specific mechanisms in the data plane for manipulating and forwarding traffic in the network so as to meet application needs. In this chapter, we explore such data plane optimization problems. We start with the basic problem of incrementally deploying SDN-enabled switches in the data plane of a traditional (non-SDN) network. Following that, we explore methods for optimizing the behavior of these switches to support various important SDN applications such as traffic engineering, coordinated scheduling, network slicing, security, fault management, and network measurement.

3.1   SDN DEPLOYMENT

The SDN deployment problem, also referred by many researchers as SDN migration and SDN upgrading, is among the most investigated problems in SDN data plane literature, and next only to the controller placement problem in the overall literature of SDN. The problem touches on the question of how to transition from a traditional network infrastructure to an SDN-enabled one. Such transition would require the network operator to deploy one or more SDN controllers, and upgrade or replace the data forwarding equipment (switches) with new one that can interface with the controller(s). While the placement and operation of the controllers was covered in the previous chapter, the deployment of the switches was left out from the discussion so far and will be covered in the following. Actually, most of the research tackling the SDN deployment problem assume the controller placement is given and therefore focus on the switch deployment aspect.

As it happens with most novel network technologies, the transition to SDN cannot happen overnight [113]. Greenfield SDN deployment in traditional networks is not practical for a variety of reasons—capital and operational costs, performance, and security risks [114]. First, replacing newly installed traditional switches is economically prohibitive. Second, switch replacement or upgrade requires coordination and planning with network operations to avoid service downtime. Third, any changes must be checked to ensure that new security issues are not introduced.

Given the above, it is reasonable to expect that the deployment of SDN switches in existing networks will happen *incrementally* and in stages by gradually replacing traditional switches with new ones over a period that spans several years. In these so-called *hybrid* SDN scenarios, traditional and SDN-enabled switches co-exist in the same network and need to work together for the routing and management of the traffic. The controllers can directly control only the flow-tables in the SDN-enabled switches, while the non-SDN network still relies on traditional protocols for routing and other operations.

There are several advantages to hybrid SDN deployment. First, hybrid SDN deployment can coincide with scheduled incremental network upgrades to avoid service impact. Second, even in such partial SDN networks, some of the benefits of using SDN can be realized. Namely, for the traffic that crosses at least one SDN switch, it is possible to apply various SDN-controlled sophisticated policies for functions such as access control, firewall actions, and other middlebox-supported in-network services [116]. Moreover, using the SDN switches it is possible to dynamically control the routing path of the flows by changing the underlying shortest path routing and thus, create a more flexible network [117]. Third, traditional routing protocols are very effective for some tasks, e.g., for robust connectivity between forwarding devices and to the SDN controllers. Hybrid SDN networks can be used complementary to them for enhanced traffic management adding flexibility to traditional routing.

The SDN switches can be either hardware-based switches or software switches (e.g., Open vSwitch – OVS) that can run on general-purpose compute platforms. On the one hand, thanks to the use of special-purpose ASICs (Application-Specific Integrated Circuits), network processors, and TCAMs (ternary content-addressable memories), hardware switches have much higher throughputs in packet processing than software switches running on general purpose CPUs. On the other hand, software switches are easier to deploy and instantiate on demand and are widely used in software-defined overlays.

A network operator wishing to incrementally deploy SDN needs a strategy for the deployment (which subset of switches to replace with SDN-enabled ones) and this has received significant attention from a modeling and optimization perspective. The focus has been on developing a good deployment strategy optimizing some performance objective or meeting some service requirement and/or cost constraint.

In the sequel, we present a basic model of optimizing SDN deployment (Section 3.1.1). Following that, we generalize the problem to capture the timing dimension in which the SDN deployment decisions are scheduled over different periods of time (Section 3.1.2). We next discuss additional deployment methods (Section 3.1.3) and conclude this section with an overview of system architectures proposed for hybrid SDN (Section 3.1.4).

3.1.1   BASIC MODEL

State-of-the-art research work models the SDN deployment problem in different ways. Below, we present one of the most basic models. As mentioned above, various services like customized monitoring, policies and access control can be applied to traffic that crosses even one SDN-enabled switch (while the rest of the switches along the routing path from origin to destination can be traditional ones). Indeed, the SDN controller can reconfigure the flow table of that single SDN switch to realize the above services in a programmable and flexible manner. Therefore, a natural question to ask is how to find the SDN deployment that maximizes the amount of such SDN-controlled traffic in the network.

We refer to the traffic that crosses at least one SDN switch as *programmable* traffic from now on. To model the SDN deployment problem, we consider a graph $G(\mathcal{V}, \mathcal{E})$ where $\mathcal{V}$ is the set of traditional switches and $\mathcal{E}$ is the set of network links interconnecting the switches. The network traffic consists of a set $\mathcal{F}$ of origin-destination flows. With traditional IP routing protocols, like OSPF, each flow f follows the shortest path to the destination. With more advanced traditional protocols, like MPLS, flow f can follow a different (non-shortest) path based on some origin-destination bandwidth declaration mechanism. In any case, we denote by $\mathcal{V}_f \subseteq \mathcal{V}$ the set containing the switches along this initial path for flow f. We further denote by λ_f (bps) the demand of flow f.

We need to emphasize that due to network dynamics the traffic demand and routing may change over time. However, once the SDN switches are deployed in certain positions, their deployment will not be changed. Thus, it is hard to find an optimal deployment scheme for frequently changing networks. To overcome this issue, it is reasonable to regard λ_f and $\mathcal{V}_f \subseteq \mathcal{V}$ as *long-term* average estimates rather than the actual current values.

The output of the problem is the number of SDN switches and their location. These decision variables can be represented as follows:

$$x_u = \begin{cases} 1 & \text{if an SDN switch is deployed at the location of traditional switch } u \\ 0 & \text{otherwise.} \end{cases} \tag{3.1}$$

The amount of programmable traffic can be expressed as a function of the deployment decisions we make, as follows:

$$J(x) = \sum_{f \in \mathcal{F}} \lambda_f \mathbb{1}_{\{\sum_{u \in \mathcal{V}_f} x_u \geq 1\}}. \tag{3.2}$$

In the above expression, $\mathbb{1}_{\{.\}}$ is the indicator function whose value is equal to one if the condition in the subscript is true, otherwise zero. Here, the condition is true when at least one SDN switch is deployed in the routing path of flow f, thereby counting this flow as programmable.

Deploying a new switch (SDN or otherwise) requires capital expenditures. These costs typically differ across locations. For example, deployment to edge nodes is typically less expensive than core network nodes. We denote with b_u the cost for deploying SDN at switch u. The network operator is assumed to have an allocated budget B for SDN deployment. The objective of the network operator is to find the deployment policy that maximizes the amount of programmable traffic subject to the limited budget:

$$\max_{x} \quad J(x) \tag{3.3}$$

$$s.t. \quad \sum_{u \in V} x_u b_u \leq B. \tag{3.4}$$

$$x_u \in \{0, 1\}, \forall \, u \in V. \tag{3.5}$$

This is a challenging combinatorial optimization problem. For example, in a network of $N = 100$ nodes there will be $2^N = 2^{100}$ possible solutions, and therefore brute force algorithms that exhaustively search the solution space will require prohibitively large running time. Fortunately, the objective function can be linearized by replacing the indicator function with an auxiliary binary variable constrained to be less or equal to the expression in the subscript of the indicator function. Therefore, the problem can be cast as an ILP and solved for small and medium size instances using standard ILP solvers like CPLEX or Gurobi.

For large-sized problem instances though it is unlikely to optimally solve the problem since it has been shown to be NP-Hard. Specifically, the problem has been shown to be equivalent to the well-known NP-Hard Budgeted Maximum Coverage Problem (BMCP) [119]. On the positive side, the BMCP problem can be approximately solved using a greedy type of algorithm. When applied to the SDN deployment problem, this algorithm iteratively deploys SDN to the switch with the highest ratio of the traffic that becomes programmable over deployment cost. At each iteration, the candidate switch set contains only the switches whose inclusion in the solution would not violate the budget constraint. To ensure the approximation ratio, the algorithm lists all triplets of switches and then for each of them greedily augments the next switch to be deployed. At the end, the best solution of all is selected.

An Illustrative Example

To better understand how SDN deployment optimization decisions are made and why the solution to this problem is not trivial, we depict the deployment decisions made by several algorithms for an example network. This network consists of 12 switches and 30 directed links as depicted in Figure 3.1 (two directed links for each undirected link are shown in the figure). The publicly available dataset for this network records the traffic matrix, i.e., the data transmitted between every pair of nodes, every 5 minutes for an overall period of 6 months. In the dataset, the traffic matrix at 8:00 pm on the first day is used to set the rates of the respective $12 \times 12 = 144$ flows (λ_f). The aggregate rate is found to be 5.46 Gbps.

The figure shows the deployment decisions made by two simple intuitive heuristic schemes: (i) DEG that deploys SDN to the switches with the highest degrees (number of incoming and outgoing adjacent links) in the topology graph, until the total monetary budget is spent; and (ii) VOL that deploys SDN to the switches with the highest traffic volume that traverses them, until the budget is spent. It also shows the deployment decisions of the greedy algorithm described above. The total budget in this example is $300K, while the deployment cost is $100K for each switch, therefore up to three SDN switches can be deployed.

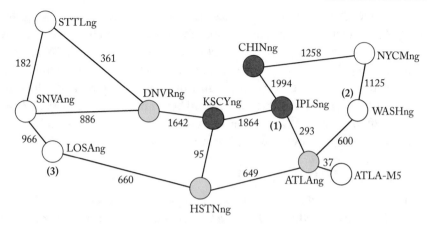

Figure 3.1: SDN deployment in the Abilene network returned by DEG (highlighted in light gray), VOL (highlighted in red), and Local Search (indexed by (1), (2), (3)) algorithms. Vertex labels correspond to switch names and link labels correspond to aggregate traffic volumes (in Mbps).

While DEG replaces with SDN the three switches with the highest degrees (ATLAng, HSTNng, and DNVRng), VOL picks the IPLSng, KSCYng, and CHINng switches which cover the most heavy-loaded links. Greedy picks first the IPLSng switch similar to VOL. However, the next decisions are different as most of the traffic that KSCYng and CHINng cover is already made programmable by the upgrade of IPLSng. Therefore, two different switches will be picked (WASHng and LOSAng). This solution of Greedy happens to be the optimal one as we can verify by exhaustively searching the solution space given the small scale of this example network.

Further Discussion

While the condition of crossing at least one SDN switch is sufficient in order to enforce policies to a flow by the controller, in some cases it makes a difference which specific switch is the SDN-enabled one. For a firewall policy, for example, of course the malicious packets can be blocked by the SDN switch (instructed by the controller) regardless of the position of the switch on the routing path of the flow. Yet, identifying these packets early, at a switch close to the origin of the flow, can expedite the handling of the problem and prevent these packets from flooding/poisoning the network further. In another example, a traffic engineering policy would benefit more if the SDN switch had many adjacent links, especially those links leading to different branches, as this would mean that this switch could dynamically re-route flows to many different paths and thereby better load balance the network. In contrast, if the SDN switch is the last one on the routing path, there would be no opportunity for the flow to be re-routed in the network since it would have already reached its destination. Generalizing the programmable traffic objective function to capture and optimize cases as the above has not been attempted so far. In addition to the location of the SDN switch, the number of SDN switches can also play a role for certain policies. For example, if we deploy SDN to two neighbor switches rather than only one of them then the two switches can work together by dynamically pushing packets to each other to respond to traffic fluctuations and better balance their loads. We also note that the timing of SDN deployments is another important dimension of this problem that we have not discussed so far. We will cover this topic in the following.

3.1.2   TIMING DIMENSION

The work in [119] suggested that the switch deployment procedure should not be in one step, but rather that a time schedule of deployments should be explicitly defined. This aspect of timing has many implications. First, like every new technology, the initial costs of new technology (such as SDN) decrease over time. Hence, there is the dilemma of early deployment that will allow the network to reap the new technology benefits immediately (e.g., make more traffic programmable right away) and a late deployment that will reduce capital expenditures (keep the budget to spend in the future). Second, the switches are highly heterogeneous since they serve a different amount of traffic and have a different remaining lifetime, and this further adds new challenges in making these decisions.

Given a total monetary budget available for a time period of multiple steps (e.g., a few years), the key question is how to spread this budget over time on SDN deployments. To answer this question, [119] generalized the programmable traffic maximization model described in Equations (3.3)–(3.5) to account for multiple time periods. Specifically, the network operator makes the SDN deployment decisions along a time interval of $t = 1, \ldots, T$ time periods, $t \in \mathcal{T}$. Typically, such decisions are made in an annual or semi-annual fashion, and by accommodating the lifetime of this type of equipment (3–5 years). Thus, a usual value can be $T = 5$ or 10 time periods. An example of incremental SDN deployment is depicted in Figure 3.2.

The generalized optimization variable $x_{tu} \in \{0, 1\}$ indicates whether SDN switch $u \in \mathcal{V}$ is deployed at time period t or not. The SDN deployment decisions must satisfy the following constraint:

$$\sum_{t \in \mathcal{T}} \sum_{u \in \mathcal{V}} x_{tu} b_{tu} \leq B, \tag{3.6}$$

where b_{tu} denotes the cost of deploying SDN switch u at time period t. These costs are expected to decrease with time, therefore it should be $b_{t'u} \leq b_{tu}$ for $t' > t$. The programmable traffic expression can be generalized as follows:

$$J^{gen}(x) = \sum_{t \in \mathcal{T}} \sum_{f \in \mathcal{F}} \lambda_{tf} \mathbb{1}_{\{\sum_{u \in \mathcal{V}_f} \sum_{t' \leq t} x_{t'u} \geq 1\}}, \tag{3.7}$$

where λ_{tf} denotes the demand of flow f at time period t. The demand is expected to increase with time, so it shall be: $\lambda_{t'f} \geq \lambda_{tf}$ for $t' > t$.

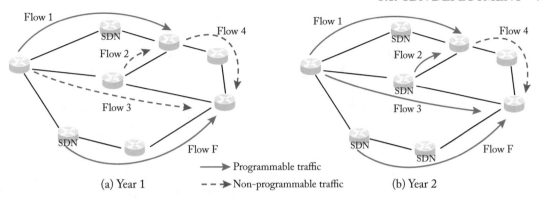

Figure 3.2: An example of incremental SDN deployment for $T = 2$ years. Two switches are deployed each year, increasing the amount of programmable traffic.

The work in [119] showed that the greedy algorithm used to approximately solve the Budgeted Maximum Coverage problem can be extended to solve the problem for multiple time periods at the cost of a logarithmic with respect to T loss on the approximation bound. To obtain a tighter approximation bound when T is large, they further proposed a local search algorithm. Algorithms of this family are quite classic in the literature of combinatorial optimization problems. They work by maintaining a feasible solution that is repeatedly updated by making "local" moves that improve the performance objective function, until no improvement is possible. For our problem, the local moves correspond to additions and deletions of a few elements (i.e., deployments/non-deployments of SDN at certain switches) in the current solution. The approximation bound of local search algorithm for this problem can be proved using submodular set function theory. The intuition is that the programmable traffic improves as the set of deployed nodes expands, hence the respective function that captures the programmable traffic benefits is monotone. However, the benefits saturate as more and more devices are deployed given that a larger portion of the traffic has already become programmable by the previously deployed devices.

Figure 3.3 demonstrates the distribution of SDN deployments (upgrades) over years when the local search algorithm is used. Here, the network operator can postpone part of the deployment decisions for a window of up to $T = 5$ years. Various scenarios are evaluated differing into the annual decrease rate of the deployment costs (b_{tu} values). We observe that for relatively low rates of cost decrease (up to 20%), all the deployments should take place within the first year. Nevertheless, after this point, it is more beneficial to postpone some of the deployment decisions in the future. The distribution of deployments over years becomes more diverse as the rate of cost decrease increases. This result highlights the complexity of the SDN deployment problem which requires considering together multiple aspects, including timing, when designing the deployment strategy.

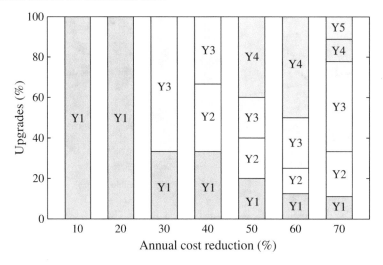

Figure 3.3: The distribution of SDN deployments (upgrades) across years (Y1, Y2, Y3, Y4, Y5) for different cost reduction rates.

3.1.3   OTHER METHODS

So far, we focused on optimizing the programmable traffic, a reasonable objective motivated by the fact that some of the SDN benefits can be realized even for a flow that crosses one SDN switch. As we briefly discussed in the end of Section 3.1.1, additional benefits may be gained for a flow crossing multiple switches, motivating the use of different objectives and deployment methods.

Consider for example the hybrid network in Figure 3.4a that needs to route a flow from source switch 1 to destination switch 3. Here, SDN is deployed to only two of the seven switches (switches 1 and 4). Considering that OSPF is the traditional routing protocol used in the network before the introduction of SDN, the flow is always routed along the shortest (with the minimum number of hops) path. However, switch 1 can dynamically decide to drop (instead of forwarding) the packets, acting, e.g., as a firewall. It can also override the default shortest path routing by routing the packets through switch 4. The packets will then follow the alternative path 1 which is the shortest path connecting switch 4 with 3. Such flow rerouting is important when a link of the shortest path fails or becomes temporarily congested. Since switch 4 is also SDN-enabled, it can similarly defer packets toward alternative path 2. In other words, as the number of SDN-enabled switches increases, the set of alternative paths increases as well. Hence, there exist more degrees of freedom (or, flexibility) in performing dynamic traffic engineering (TE). Such scenarios are clearly beyond the definition of programmable traffic and necessitate the adoption of novel models and problem formulations.

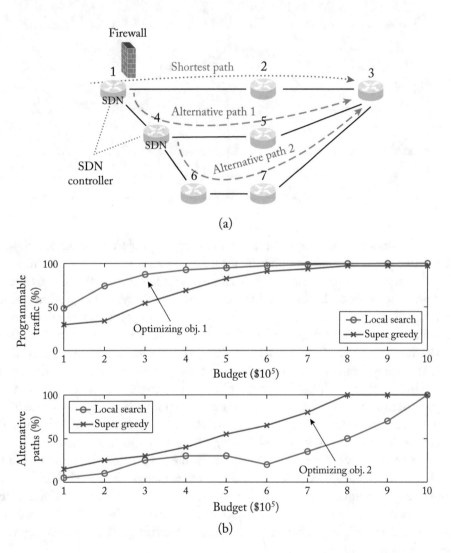

Figure 3.4: (a) Dynamic control of the routing path by two SDN-enabled switches. (b) Impact of optimizing different objectives.

Table 3.1: State-of-the-art work on SDN deployment

Ref.	Objective			Timing	Joint with Routing
	Programmable Traffic	TE	Cost		
[119, 120]	✓	✓	✓	✓	✗
[121]	✓	✗	✓	✗	✗
[122]	✓	✗	✓	✗	✗
[123]	✓	✗	✗	✗	✗
[124, 125]	✗	✓	✗	✗	✗
[126, 127]	✗	✓	✗	✓	✗
[128]	✗	✓	✗	✗	✓
[129, 130]	✗	✓	✗	✗	✓

The work in [119] modeled the problem for the objective of maximizing the number of alternative paths that are dynamically available through the SDN deployment. They showed that this objective, referred to as TE flexibility, does not exhibit the property of submodularity that programmable traffic does, and therefore local search type of algorithms would not be suitable for this objective. They proposed instead a different solution technique that quantifies the so-called supermodular degree of the function, i.e., how much the function deviates from the submodular property. Based on such analysis, another variant of the greedy algorithm (called "Super-Greedy") can be shown to achieve approximation guarantees for this problem.

The Local Search algorithm we discussed in the previous section is in practice a very efficient algorithm for maximizing programmable traffic. Yet, a large volume of programmable traffic cannot guarantee by itself a large number of alternative routing paths (and vice versa). Therefore, it is questionable how well an algorithm that optimizes one of the two objectives will perform with respect to the other objective. Figure 3.4b, originally presented in [119], aims to shed light on this issue by comparing the performance of the Local Search algorithm (which optimizes programmable traffic) and Super-Greedy (which optimizes TE flexibility). The results show that by optimizing one of the objectives, benefits are also realized for the other objective. However, there will be a performance loss (up to a factor of 2), since each algorithm favors one objective over the other.

Other recent works have studied the SDN deployment problem. Table 3.1 lists the main works in the literature categorized based on their objective—whether they consider the timing aspect of the problem or not, and whether they optimize traffic routing together with the deployment decisions or not. We will discuss the latter issue more in the sequel. Most of the works formulate the problem as maximizing performance (traffic programmability or TE benefits) subject to a deployment cost (budget) constraint while others consider the analogous version of minimizing cost subject to a performance guarantee constraint. Reference [120] extends [119] by proposing a partial cover type of algorithm for minimizing the deployment cost while achieving a lower bound on the programmable traffic amount. This algorithm performs a Lagrangian relaxation of the problem that can be approximately solved using another greedy algorithm. Then, the method performs a binary search on the Lagrangian value to output the final solution. References [121] and [122] proposed greedy-based heuristics for the programmable traffic maximization and cost minimization objective, respectively. Reference [123] designed a cover-based approximation algorithm for the programmable traffic maximization objective. A distinctive feature of the latter work is that the SDN devices do not replace the traditional ones, but are deployed in addition to them. On the one hand, this "duplicate" approach does not change the traditional network, and therefore it does not impose any operational risks from SDN software faults or attacks. On the other hand, this approach requires additional cost for network connections between the SDN and traditional switches. This network connection cost can be small though if the SDN devices are located near to the traditional devices. In another front, reference [124] studied the objective of maximizing the minimum number of loop-free routing paths enabled by the SDN switches. This TE type of objective can have a positive impact by eliminating the routing bottlenecks likely to be utilized by an adversary. Another approach that deploys SDN switches in a way that partitions the network into sub-domains so as to achieve TE capabilities comparable to full SDN deployment was proposed in [125]. However, these last works did not consider the timing aspect of the deployment problem. This aspect was considered by references [126, 127]. Yet, unlike [119, 120], these works did not provide any approximation bounds nor analyze the impact of equipment cost reduction over time.

Joint Deployment and Routing

Another series of works [128–130] follow a joint SDN deployment and routing optimization approach, i.e., the flow routing and switch deployment decisions are jointly made. This is different from all the previous work discussed above that took the flow routing paths as fixed, determined by the traditional routing protocols. The motivation though is not the same among these three works. On the one hand, [128] is based on the simple observation that if we are permitted to optimize the routing of traffic flows in the network then we can carry more of it through the SDN switches we deploy thereby increasing the programmable traffic. The joint deployment and routing problem is then formulated to maximize the amount of programmable traffic. Evaluation results indicate that when 10% of the SDN switches are deployed among traditional switches then about 25% of the traffic is made programmable.

On the other hand, the motivation of [129, 130] is that the SDN switches once deployed will be able to route traffic to any of the next hop switches by overriding the traditional protocol, with the location of the SDN switches being the ones determining the available routing options. Therefore, by jointly optimizing the location of switches and the traffic routing we can achieve the best possible routing performance (e.g., maximum throughput or minimum maximum link utilization). Although the above joint approaches are more advanced and can boost network performance since they leverage more degrees of freedom when making the optimization decisions, they face one major challenge. While being obviously related to one another, the SDN deployment and traffic routing actions are taken on different timescales. The SDN switches locations are fixed once deployed, while the routing paths should change frequently to respond to the network dynamics. Therefore, even if routing is optimized together with the deployment, it will have to be re-optimized again in future for the given deployment every time traffic demands change. Actually, this is the reason that all the other previous work considered these two problems separately. We elaborate in the next section (Section 3.2) on the problem of optimizing routing for a given SDN deployment.

3.1.4   HYBRID SDN ARCHITECTURES

For completeness, we note that there exist various architectural choices for deploying hybrid SDN. Regardless of the architecture chosen, the SDN switches can deviate from the decisions of the traditional protocols for the packets traversing them to enable more flexible routing and control and, therefore, all the methods presented so far can be used to optimize SDN deployment. Specifically, there exist the following four main architectures of hybrid SDN illustrated in Figure 3.5.

- **Topology-based.** The network is divided into areas, also called islands or zones. All the switches in the same area are controlled by the same paradigm: SDN or non-SDN. For example, an operator may use SDN for its data center network while using traditional protocols in its WAN. We note that this approach can be a good fit for existing multi-domain networks where SDN can be deployed in only some of the domains. The SDN controller can be used in these networks to help realize inter-domain services by interfacing with a traditional inter-domain protocol such as BGP.

- **Service-based.** In this architecture, SDN and traditional protocols provide different services. SDN can be used to provide services that benefit from fully centralized traffic monitoring or centralized enforcement of traffic engineering policies, while other services are handled by traditional protocols. Unlike the topology-based architecture, the two approaches can co-exist over an overlapping set of switches, controlling a different portion of the traffic and forwarding table of each switch. This is possible with switches that support both approaches. Such switches can be traditional switches that also support SDN functionality, and in some cases software switches or new hardware switches supporting such hybrid operation.

- **Class-based.** This architecture partitions the traffic into classes and each class can be controlled either by SDN or traditional means. In contrast with the service-based architecture, each paradigm realizes all the network services for the traffic classes assigned to it. For example, we can classify flows to implement fine-grained, load-balancing for those flows attracting most traffic and to guarantee low delay for flows corresponding to interactive applications. Similar to

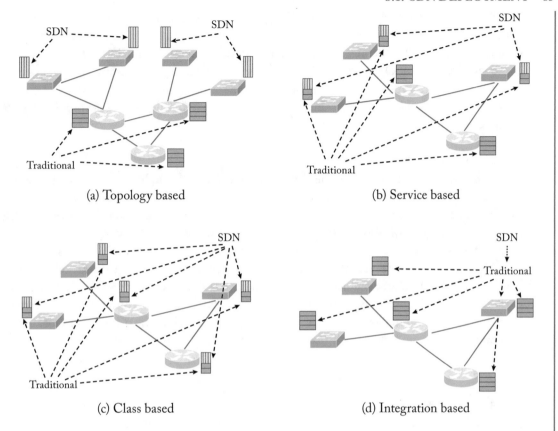

Figure 3.5: Hybrid SDN architectures.

the service-based architecture, this architecture also requires hybrid switches supporting both SDN and traditional control.

• **Integration-based.** In this architecture (also referred to as controller-based), SDN is used to control the operation of traditional protocols through adjusting the protocol parameters, e.g., the OSPF weights. This is done using an appropriate southbound interface that allows the SDN controller to interact with the traditional switches. Note here that this interaction is only with the control plane of the switches. The actual packet forwarding tables are assumed to be controlled only by the traditional protocols. This is different from the previous architectures where SDN and traditional protocols complemented each other controlling different parts of the forwarding tables of the switches.

Fibbing [131] is an example of the integration-based architecture. Fibbing introduces fake switches and links into the underlying link state routing protocol to indirectly make it change its decisions and manipulate it so as to achieve some level of load balancing and traffic engineering.

A limitation of Fibbing is that its forwarding rule matching is limited to destination-based, and its expressivity is thus the same as the expressiveness of IP routing. Besides, with injected "lies," Fibbing could lead to debugging issues and incorrect operation. DEFO (Declarative and Expressive Forwarding Optimizer) [132] is another architecture in this family. It leverages segment routing to control routing paths for carrier-grade traffic engineering but has some constraints as in Fibbing. Besides, its constraint programming based middle-point selection largely focuses on static traffic matrices.

Further Discussion

When deploying SDN, first is the choice of architecture for hybrid SDN, second the type of switches, and third the optimization strategy to determine the actual number and location of SDN switches. A right choice of architecture depends on the needs and priorities of the network operator, the types of traffic and applications. For example, if traffic congestion events are observed to occur more often only in a certain area of the network, it would make sense to deploy SDN to this area only to handle traffic fluctuations following a topology based architecture. Service and class based architectures should be preferred when there is a wide range of services and traffic types with diverse requirements, and so SDN can be used to control the most demanding of them. An integrated-based approach may be preferred when the priority is to make minimal changes to networks and mostly use the well-established traditional protocols.

3.2 TRAFFIC ENGINEERING

Traffic Engineering (TE) refers to strategies and algorithms for routing traffic in a network so that traffic demands are met while a performance objective is optimized. This usually requires the computation of optimized paths between source and destination nodes in the network. Various objectives can be considered depending on the type of network and application such as minimizing traffic delay by selecting shortest paths (e.g., paths with minimum number of hops), minimizing congestion in the network by splitting traffic across multiple paths to balance link loads, and minimizing packet loss by overprovisioning resources of multiple paths, and enabling a fast-reroute mechanism upon detection of path failure. In this section, we give an overview of such TE methods in SDN literature.

3.2.1 HOW CAN SDN HELP?

Traditionally, networks relied on distributed mechanisms to perform TE. Shortest Path First (SPF) algorithms, in use since the early days of the Internet, can minimize path length and achieve low end-to-end delays (a critical metric for real-time communications). However, this approach leads to unbalanced usage of network resources where shortest paths can end up congested while other paths are underutilized. To mitigate this problem, the idea of traffic splitting was introduced where traffic was split equally among all available shortest paths, known as the Equal Cost Multipath Protocol (ECMP). Later on, Multi-protocol label switching (MPLS) protocol enabled increases in performance and flexibility for routing. The main idea behind MPLS is to make packet forwarding decisions based on previously-assigned labels (short bit sequences) and flows can be sent along highly efficient pre-computed paths. This allows for differentiation in routing for packets from applications with different quality-of-service needs. For example, packets carrying real-time traffic, such as voice or video, can take different labels so as to be easily mapped to low-latency routes across the network—something that is challenging with shortest path based routing mechanisms.

SDN provides a number of appealing features to improve the performance and flexibility of TE. First, in SDN the path computation can be performed in the **logically centralized control plane** which is aware of the network state in real time. Use of accurate real-time state information in path computation leads to better paths and performance. Second, the **higher granularity** in routing based on different packet headers (not just destination header) can be used for more efficient traffic splitting ratios and better flow dissagregation. Third, the **programmability** provided by the SDN controllers, which are fully configurable entities, facilitates the implementation of various features such as TE with service chaining and various dynamic, fast failure recovery mechanisms.

The above capabilities of SDN spurred a new wave of research on TE. While for many years researchers have been striving to come up with routing mechanisms that are distributed, lightweight and easy to implement, SDN significantly changed the design criteria and priorities. The logically centralized control plane of SDN shifted the attention to centralized routing algorithms, where centralized execution is no longer considered a problem but centralized algorithms can be easily implemented in real networks. Moreover, since SDN is expected to be deployed gradually and thus co-exist with traditional network devices (as we discussed in the previous section), new TE problems were defined in literature tailored to the distinct characteristics of these hybrid scenarios. In the following, we describe such a representative model of TE. We next discuss online optimization TE methods, practical constraints on memory, and energy costs, and conclude this section with advanced methods based on segment routing.

3.2.2 REPRESENTATIVE MODEL

Most TE problems in full SDN networks (all switches are SDN-enabled) are equivalent to the multicommodity flow problems [133], which were well studied years ago—before SDN appeared. Problems in this family aim to select routing paths for flows of different types (commodities) in a network such that congestion is avoided. This can be achieved by minimizing the maximum link utilization (MLU), a widely used metric for load balancing the links in a network. Below, we present a model of the multicommodity flow problem.

Given a network represented by the graph $\mathcal{G}(\mathcal{V}, \mathcal{E})$ with $\mathcal{V}$ switches and $\mathcal{E}$ links, the capacity of link e is denoted by c_e and the set of all flows in the network by $\mathcal{F}$. Each flow f has demand λ_f (bps) and can be routed through a set of possible paths $\mathcal{P}_f$. In the unsplittable (or integral) routing version of the problem, each flow has to be routed without being split through only one of the paths. In the splittable (or fractional) case, the demand of a flow can be split and routed through multiple paths. This is also known as multipath routing. On the one hand, unsplittable routing is more challenging since it leads to intractable integer program formulations. On the other hand, splittable routing can be modeled as a linear program that can be optimally solved in polynomial time. Besides, by allowing to split the flows over multiple paths, the link loads can be better balanced, though at the risk of delivering packets of the same flow out of order.

We define the problem for the splittable routing case. A real variable y_{fp} denotes the portion of the flow f that selects the path $p \in \mathcal{P}_f$. Then, the problem of minimizing the MLU can be expressed as follows.

Multicommodity Flow Problem: $\qquad \min \theta \qquad\qquad\qquad\qquad$ (3.8)

$$s.t. \sum_{p \in P_f} y_{fp} = 1, \ \forall f \in \mathcal{F} \qquad\qquad (3.9)$$

$$\sum_{f, p \in P_f | e \in p} y_{fp} \lambda_f \leq \theta c_e, \ \forall e \in \mathcal{E} \qquad (3.10)$$

$$y_{fp} \in [0, 1], \ \forall f \in \mathcal{F}, p \in \mathcal{P}_f. \qquad (3.11)$$

The first set of equations requires all the demand of a flow f to be routed. The second set of inequalities ensures that the total flow volume on a link is less than the product of the variable θ and the capacity of the link (c_e). The last set of constraints requires the flow splitting portions to be between 0 and 1. Therefore, θ takes the value of the MLU in the optimal solution of the problem. Note that if $\theta < 1$, then none of the links is over-utilized. However, if $\theta > 1$ it means that some of the links suffer from congestion.

As discussed in the previous chapter, the SDN controller collects statistics about the network and is aware of the flow demands and available routing paths. Therefore, it can centrally solve the problem as a linear program and find the values of the variables (θ and y_{fp}). The controller can opt to solve the problem periodically (e.g., every few minutes or hours) or in an event driven manner (e.g., every time it detects that link loads have become imbalanced due to high network dynamics). The y_{fp} values will be translated to packet matching rules installed by the controller at the flow tables of the switches.

Hybrid SDN

We next describe how the multicommodity flow problem needs to be revisited to address the case of hybrid SDN. This problem was studied for the first time in [134]. In hybrid SDN, only a subset of the switches are SDN-enabled while the rest are traditional switches running a protocol like OSPF that assigns routes to shortest paths. Formally, let $\mathcal{S} \subseteq \mathcal{V}$ denote the subset of SDN-enabled switches and $\mathcal{S}^c = \mathcal{V} \setminus \mathcal{S}$ be the non-SDN-enabled (traditional) switches.

For a traditional switch $u \in S^c$ the next hop switch to forward the packets of a flow is fixed and follows the OSPF protocol. Specifically, we denote by $NH(u, d)$ the next hop switch in the shortest path starting from switch u to switch d. For ease of explanation, we assume that there is only a single shortest path and so a single next hop switch for each pair of switches u and d. Therefore, $NH(u, d)$ contains a single switch rather than a set of switches. For an SDN-enabled switch $u \in S$, the controller can pick the next hop switch freely and overwrite the next hop switch decision of the traditional protocol.

Since the only traffic that the SDN controller can manipulate is the traffic that passes through the SDN-enabled switches, we just focus on the traffic injected at these switches. We say that an SDN-enabled switch $u \in S$ **injects** a packet if u is on the OSPF routing path for the packet and the packet passes through u before any other SDN-enabled switch. The total traffic injected by switch u that is destined to some switch d is denoted by I_{ud}. The traffic not passing through any of the SDN switches is not programmable and cannot be manipulated by the controller. We denote by g_e the non-programmable traffic that traverses link e. In practice, the controller can estimate the values of I_{ud} and g_e based on measurements made by the SDN-enabled switches and the OSPF messages received by the controller.

For each unit of traffic I_{ud} injected by a switch u, the controller can find one or more alternative paths for routing. We define by $\mathcal{P}_{ud}$ the set of these paths between switches u and d which we call **admissible** paths from now on. Formally, a path $u = v_0, v_1, v_2 \ldots, v_k = d$ is called admissible if for all $j = 1, 2, \ldots, k$, we have that $(v_{j-1}, v_j) \in \mathcal{E}$ and $v_j = NH(v_{j-1}), d$ if $v_{j-1} \in S^c$, which is just a mathematical way to represent that the next hop routing decisions of the traditional switches in S^c are determined by the OSPF protocol. The notation $\mathcal{P}$ denotes the union of all these paths.

The objective of the controller is to route the traffic injected by the SDN switches through admissible paths such that the MLU is minimized, which can be formulated as follows.

Multicommodity Flow Problem in Hybrid SDN:

$$\min \theta \tag{3.12}$$

$$s.t. \sum_{p \in \mathcal{P}_{ud}} y_p \geq I_{ud}, \ \forall u \in S, d \in V \tag{3.13}$$

$$g_e + \sum_{p \in \mathcal{P} \mid e \in p} y_p \leq \theta c_e, \ \forall e \in E \tag{3.14}$$

$$y_p \geq 0, \ \forall p \in \mathcal{P}. \tag{3.15}$$

In the formulation, the variables are y_p, which is the flow demand in admissible path p. Note how this is different from the notation y_{fp} in the previous formulation in Equations (3.8)–(3.10) since this time p is taken over the paths between pairs of switches u and d and therefore we do not need to denote separately the decisions for each flow f. The first set of inequalities ensures that the total injected traffic is routed in the network through some admissible paths.

The second set of inequalities ensures that the total flow on the link which is the sum of the unprogrammable flow (represented by g_e) and the programmable flow is less than the product of the MLU (θ) and the capacity of the link (c_e). The third set of inequalities ensures that the flow on any path is non-negative. The above is a linear program and so it can be optimally solved by the SDN controller in polynomial time using standard linear optimization techniques.

3.2.3   ONLINE OPTIMIZATION

The formulations in the previous section are offline meaning that the controller knows the demands of the flows in the network and therefore can select routes for all of them to optimize performance. In the online version of the problem, traffic flows arrive into the network one at a time and have to be routed to their destinations straightaway without knowing the future flow arrivals. A newly arrived flow can use only the bandwidth left available by the ongoing flows. If there is not any path with enough bandwidth in the network the flow cannot be admitted and therefore the overall throughput of the network will be affected. Maximizing throughput in this case is an important objective just like the MLU minimization we considered in the previous section.

We model the problem for the case where SDN is fully deployed in the network just like we did in the first formulation in the previous section. Extension for the hybrid case is rather straightforward and can be found in [134]. We denote by f_i the i-th flow arriving in an online fashion that consists of a source switch s_i, a destination switch d_i, and a set of feasible routing paths P_i that can be used to serve that flow. The set P_i is assumed to be given and may contain a number of shortest paths (e.g., with respect to the hop count length) connecting the source and destination of that flow. For simplicity, we assume that all flows require one unit of bandwidth.

There exist several variants of the online routing problem. Here, we consider the one in which (i) flows can be split along several paths (splittable) and (ii) once a flow is routed, no rerouting is allowed (non-preemptive). As we discussed already in the previous section, splittable routing can improve performance compared to the unsplittable case where the flow has to follow a single path. On the other hand, preemptive routing can also improve performance but it brings signaling overheads and additional latency for re-routing during preemption. The optimal routing problem can be formulated as the following online multicommodity flow type of problem.

Online Multicommodity Flow Problem:

$$\max \sum_i \sum_{p \in P_i} y_{ip} \tag{3.16}$$

$$s.t. \sum_{p \in P_i} y_{ip} \leq 1, \ \forall i \tag{3.17}$$

$$\sum_{i,p \in P_i | e \in p} y_{ip} \leq c_e, \ \forall e. \tag{3.18}$$

In the above formulation, the variable y_{ip} indicates the amount of bandwidth allocated to flow f_i on path p. The objective is to maximize network throughput. The objective of minimizing the MLU can be also modeled similar to the previous section. The first set of constraints guarantees that the total bandwidth allocated to each request is at most 1. The second set of constraints guarantees that the edge capacities are not violated.

Since the flows arrive in an online fashion, we cannot solve the problem for all flows. To find an online solution that makes decisions for each flow at the time of its arrival, we can fit the above problem into a primal-dual framework, as discussed in [135]. Specifically, we can regard the above linear program as the "dual problem" and define the "primal problem" as follows.

$$\textbf{Primal Problem:} \qquad \min \sum_e c_e A_e + \sum_i B_i \qquad (3.19)$$

$$s.t. \sum_{e \in p} A_e + B_i \geq 1, \ \forall i, p \in P_i, \qquad (3.20)$$

where A_e and B_i are the variables of the primal problem.

Primal-dual algorithms typically maintain a solution to the primal program as they run that upper bounds the value of the current dual solution. The work in [135] proposed such a primal-dual algorithm and formally proved that it is $O(\log |\mathcal{V}|)$ competitive. This means that the algorithm achieves throughput at least $1/\log |\mathcal{V}|$ of the maximum possible one that would have been achieved by an offline algorithm that has full knowledge of all the flow arrivals. For each newly arrived request, the primal-dual algorithm routes it if there is a path of "cost" less than 1, thereby ensuring that the cost of routing this flow will not be more than the increase in the objective function (throughput). The interesting point is how the cost is defined. Specifically, the algorithm will define the cost of a path as the sum of the A_e variable values of the primal problem. Each time a new flow is routed by the algorithm, these variables will be updated so that the cost increases by an exponential factor with respect to the total number of switches in the network for all the links in that path. This way, the future requests will be discouraged from picking links in that routing path.

The above primal-dual algorithm can be easily extended to handle flows with different utilities and bandwidth demands, as well as for other objectives such as minimizing the MLU or maximizing fairness. A max-min fair routing solution for example is an allocation of bandwidth to flows that one cannot increase the bandwidth allocation to a flow f_i without decreasing the bandwidth allocated to flows that have received at most the bandwidth given to f_i.

We remark that further extensions of the primal-dual algorithm have been proposed for more complex problems such as the online bulk transfers in Inter-Datacenter WANs subject to multiple deadline, budget, and bandwidth constraints [136], as well as the online joint routing, composition, and placement of virtual network functions [137]. The work in [138] suggested an algorithm that shares some similar ideas with the primal-dual. It considered the online multicast problem where traffic needs to be routed from a source to multiple destinations. The authors proposed an algorithm that selects low-cost multicast trees where the cost is an exponential function of the available bandwidth of the links in the tree, similar to the way the primal-dual algorithm works.

We also remark that different types of online algorithms, beyond the primal dual, might be also promising to address the online routing problem. For example, online gradient/mirror descent methods have been shown very effective for problems in which instantaneous decisions are constrained to lie on a fixed convex set. However, these methods cannot typically capture long-term constraints coupling decisions taken in different times such as the constraint in Equation (3.18). A few first attempts to extend these algorithms for long-term constraints have been made, e.g., see the recent work in [139], but the area is still open for future research.

3.2.4   ADDITONAL ROUTING CONSIDERATIONS

We consider two auxiliary costs associated with routing that have been considered in the literature. The first is limited TCAM capacities for flow entries—this is mainly motivated by low-end switches. The second is energy consumption minimization motivated by the increasing energy consumption in data centers.

Memory Cost

To realize the routing strategies described above, flow rules need to be installed by the controller in each SDN switch. These rules are usually stored in the switch's ternary content-addressable memory (TCAM) which can perform parallel packet matching at high speed. Unfortunately, this efficient memory is relatively expensive and thus limits the maximum size of the table particularly in low-end switches. Also, SDN rules can be more complex than classical destination-based routing rules in traditional IP forwarding. For instance, an SDN rule may match packets on source, destination IP addresses, port numbers and protocol to realize a sophisticated access-control application. The TCAM of many switches can support only a limited number of these complex rules. In contrast, there may be a huge number of flows going through the same switch making it impossible to use a dedicated flow rule to process the packets of each flow. There have been two main ways to address the limited TCAM size problem.

1. Select the routing paths under constraints on the TCAM size. This represents another criterion for making routing decisions in addition to the path length and the balancing of link loads. It may mean that a flow should be routed through a longer path than the shortest one and/or a path that is not the least loaded one, as long as the switches in the selected path have enough storage space.

2. Reduce the space in the TCAM taken by the flow rules by compression techniques. The most common way of compression is by applying wildcard rules that aggregate multiple rules that have some common matching header field(s), such as common source IP address, into one rule.

For the first way, the works in [141, 142] augmented memory constraints to the multi-commodity flow problem. These constraints are per switch and can take the following form:

$$\sum_{f \in \mathcal{F}} \sum_{p \in \mathcal{P}_f \mid u \in p} 1_{\{y_{fp} > 0\}} \leq b_u, \ \forall u \in \mathcal{V}, \tag{3.21}$$

where b_u is the memory capacity of switch u and $1_{\{.\}}$ is the indicator function; it is equal to 1 if the condition to the subscript is true, otherwise 0. This means that each flow crossing switch u contributes one rule to the memory of that switch regardless of the portion of the flow routed. This is a nonlinear function that complicates the problem. The multicommodity flow problem is not a linear program any more and cannot be solved with standard linear optimization techniques. The work in [141] proposed a randomized rounding algorithm to approximate the solution to this problem. Algorithms of this family are broad in scope and have been also used to address the hybrid SDN deployment problem discussed in the previous section, as well as network slicing problems we will discuss in the sequel. The algorithm in [141] starts with solving a linear relaxation of the problem where the above constraint is replaced with:

$$\sum_{f \in \mathcal{F}} \sum_{p \in \mathcal{P}_f \mid u \in p} y_{fp} \leq b_u, \ \forall u \in \mathcal{V}. \tag{3.22}$$

Then, independently for each flow $f \in \mathcal{F}$ and each path $p \in \mathcal{P}_f$, the algorithm chooses this path with probability y_{fp}. Finally, for each chosen path $p \in \mathcal{P}_f$, it rounds over p a portion of the flow f equal to $1/6 \log(|\mathcal{V}|)$.

Note that the solution to the linear relaxed problem in the first step of the algorithm may not be feasible, i.e., the memory capacity of the switches may be violated. The violation factor can be huge since in an extreme case the relaxed solution could route a small portion of each flow through the same switch hence installing to it the maximum possible number of flow rules. The next two steps of the algorithm will pick a single path for each flow so that the memory capacity violation factor becomes smaller and the solution is not trivial. To achieve this, the algorithm uses the power of randomization. Specifically, by randomly picking paths with probabilities equal to the relaxed solution values and then assigning demand to them scaled down by a logarithmic factor $(1/6 \log(|\mathcal{V}|))$, the authors in [141] formally prove that the memory constraints will not be violated by more than a factor of $O(\log |\mathcal{V}|)$, link bandwidth capacities will only be violated with negligible probability, and finally the total throughput value will be at least a fraction $1/O(\log |\mathcal{V}|)$ of the optimal value. Note how here the authors chose to maximize the throughput rather than minimize the MLU. An interesting open question is whether the theoretical analysis and bounds can be extended for the MLU objective.

The work in [142] proposed a similar problem formulation with memory capacity constraints and considered two objectives; minimizing the total memory usage and maximizing traffic satisfaction. The latter is just a weighted sum of the throughput demands. A novel aspect of this work is the consideration of a default rule at each switch forwarding the packets to the

SDN controller. This rule has low priority and will be used only in the case that no other rule exists to match the packets of a flow and forward them to the corresponding paths. The authors observe that if the next hop switch to a routing path chosen is the same as in the default path then there is no need to install an extra rule at this switch for this flow. Therefore, the constraint in Equation (3.22) can be modified so that for this case it does not increase the memory occupancy by an extra rule. A heuristic algorithm is proposed to solve the problem. To save memory space, the algorithm routes packets through their default paths until some deflection switch after which the packets follow the shortest path to their destination. This idea is interesting and worth further investigation to understand if theoretical results beyond the heuristic solutions can be shown.

For the second way, the work in [140] suggested to compress the flow rules by aggregating them based on their source and destination addresses. Unlike the previous works that changed the routing paths to take advantage of the flow table space of many switches, rule compression keeps the same paths and therefore it does not impact the QoS of the network. The authors defined a flow rule as a triplet (s, t, p) where s is the source of the flow, t its destination, and p the output port in the network. To aggregate the different rules, they used wildcard rules that can merge rules by source (i.e., $(s, *, p)$), by destination (i.e., $(*, t, p)$) or both (i.e., $(*, *, p)$, the default rule). A rule of the form $(s, *, p)$ means that all packets with source s and output port p will match in that rule regardless of their destination. For the online routing case, they proposed a heuristic algorithm that pushes newly arrived flows to paths with low link capacity utilization and memory space occupied. In a second stage, the algorithm greedily and repeatedly replaces rules with the same source and port but different destinations with a wildcard rule of the form $(s, *, p)$. Similar versions of the algorithm for the wildcard rules of the form $(*, t, p)$ and $(*, *, p)$ were discussed.

Energy Cost

Reducing energy consumption has become a key topic of networking research due to the ever-increasing power consumption due to large network growth and especially in data centers. The use of energy-aware routing strategies has been shown as very effective in reducing the energy spent in a network with significant gains of up to 50% reported in data center networks [143]. Since energy consumption of a switch has components that are largely independent from its traffic load, researchers have looked at ways to put switches to sleep in order to save energy. Several works have proposed routing algorithms that drive traffic away from links so they and their adjacent switches can be shut down while satisfying QoS constraints. On the one hand, the capabilities provided by SDN can be useful to facilitate the implementation of such energy-aware routing algorithms. On the other hand, the TCAMs used in many high-speed SDN switches have high energy consumption and large footprint [144].

One of the most popular energy-aware routing solutions for data center networks is the so-called ElasticTree [143] which is implemented on a testbed consisting of Openflow switches.

The authors defined a standard multicommodity flow routing problem, as the one we described above. To optimize for energy, they added to the problem binary variables for every link and switch, and constrain traffic to only traverse active (powered on) links and switches. This constraint can be expressed as follows:

$$\sum_{f \in \mathcal{F}} \sum_{p \in \mathcal{P}_f \mid e \in p} y_{fp} \leq c_e a_e, \ \forall f \in \mathcal{F}, e \in \mathcal{E}, \tag{3.23}$$

where the variable a_e indicated whether link e is active ($a_e = 1$) or not ($a_e = 0$).

It is worth mentioning that the power model used in this work ensures that the full energy cost for an Ethernet link is incurred when either side is transmitting; there is no such thing as a half-on Ethernet link. The objective of the problem is to minimize the total energy consumption, while satisfying all constraints (link capacity, flow conservation, demand satisfaction, traffic to only active links). A greedy bin packing type of algorithm is proposed that tends to pick the routing paths that are the leftmost available in the fat tree data center topology. This way more paths remain without any traffic and hence there are more opportunities to deactivate links, while the specific topology characteristics of the data center network are exploited.

3.2.5   SEGMENT ROUTING

SDN can be combined with the Segment Routing (SR) paradigm to enable much more powerful TE solutions. SR is a source routing protocol proposed by IETF. The key idea behind it is that a source node (i.e., network switch) can specify a path as an ordered list of segments to guide a packet across the network. A segment is an instruction a node executes on the incoming packet. This instruction can be in fact be the address of an intermediate node that the packet needs to traverse before reaching its destination. This way a routing path from source to destination is divided into parts; from one intermediate node to the next one in the ordered list, and so on until the packet reaches its destination. SR can be implemented independently from SDN, for example with the help of the MPLS label switching mechanism. However, the SDN controller can be useful in centrally computing the segments and distributing the respective ordered lists to the nodes.

The routing from one intermediate node to the next is done using a traditional (non-SDN) distributed protocol, like shortest path routing. Therefore, SR is a hybrid approach that combines both centralized and distributed control paradigms. The rationale is that this way we can achieve most of the benefits of SDN in terms of TE performance in a simple and scalable manner by re-using existing distributed protocols. The intermediate nodes along the path do not need to maintain state for the paths through the network since all the information needed for the routing is included in the packet header.

To better understand how SR works, we consider the example in Figure 3.6 [151]. The source node i adds a set of labels to the IP header of the packet with the address of the intermediate nodes the packet needs to traverse (m, l, k, j nodes in the example). These labels are used as

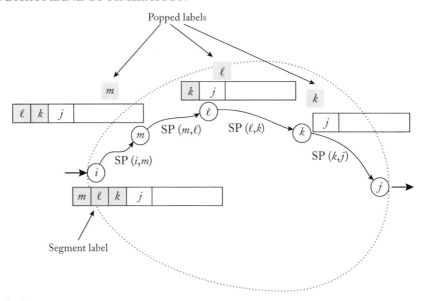

Figure 3.6: Illustration of Segment Routing. Segment labels inside the packet determine the routing of packet from node i through intermediate nodes m, l, k, and j.

temporary destination addresses for the segment. The packet is then routed using the standard shortest path routing algorithm. When the packet reaches the intermediate destination from the source, the top level label is *popped* by the intermediate destination and now the packet is routed from the intermediate node to the end point of the next segment again along the shortest path.

The TE optimization problem previously defined as a multicommodity flow problem can be revisited for the case of segment routing. The difference is that instead of defining a variable y_{fp} for each flow f and path p, we now define the variables per intermediate node. Specifically, for the simple case that only one intermediate node is permitted for each flow, a case referred to as 2-SR in [151], we define the variable y_f^k to indicate the demand of the flow f that is routed through intermediate node k. Then, the TE problem that minimizes the MLU can be formulated as follows.

$$\textbf{2-SR Problem:} \qquad \min \ \theta \tag{3.24}$$

$$s.t. \ \sum_{k \in \mathcal{V}} y_f^k \geq \lambda_f, \ \forall f \in \mathcal{F} \tag{3.25}$$

$$\sum_{f \in \mathcal{F}} \sum_{k \in \mathcal{V}} g_f^k(e) y_f^k \leq \theta c_e, \forall e \in \mathcal{E} \tag{3.26}$$

$$y_f^k \geq 0, \forall f \in \mathcal{F}, \tag{3.27}$$

where $g_f^w(e)$ is the fraction of the flow f that traverses link e through intermediate node k. The value of $g_f^k(e)$ can be computed efficiently for a given topology and link weights used by the shortest path protocol assuming that the traffic matrix is known. Constraint (3.25) ensures that all the demand is routed, constraint (3.26) is the link capacity constraint and determines the value of θ, and constraint (3.26) ensures that the flow on any segment is non-negative. This is a linear program and hence it can be easily solved using standard linear optimization techniques.

Since often the traffic matrices are not known in advance, the authors in [151] addressed the problem for two more realistic scenarios: (i) the traffic-oblivious scenario where $g_f^k(e)$ values are not known in advance and (ii) an online scenario where we know the current link loads but not any future flow arrivals. For the traffic-oblivious scenario they proposed an adversarial approach based on game theoretic techniques. For the online scenario, they proposed a simple algorithm that admits and routes flow requests based on a set of dual variable values. Evaluation results demonstrated that these two algorithms work well in practice and that 2-SR suffices to obtain most of the benefits of SR.

The SR problem has been also studied for other objectives besides MLU such as energy consumption minimization and the minimization of the number of rejected requests. More recent work considered also the overheads of SR both in terms of bandwidth use due to the insertion of the SR header in the packets, and the number of SR steering policies to be configured in the nodes. An extensive survey and classification of work in this area can be found in [152].

3.3   SCHEDULING AND TRAFFIC CONTROL

In the previous section, we described how SDN can be leveraged to empower advanced TE mechanisms by performing the computation of routing paths in a centralized manner, at the SDN controller, rather than in a distributed manner in which decisions are made by the switches hop-by-hop (e.g., OSPF protocol). Through an appropriate routing path computation, the controller can spread the flows across different paths to balance the loads in the network and prevent congestion. As the demand of the flows can change from time to time, the controller should periodically re-compute the routing paths and send the updated forwarding rules to the switches, a process that typically happens in coarse timescales of minutes or more.

While the centralized control of the routing paths can be effective in preventing network congestion, it does not address the problem completely. For instance, in a massive traffic demand scenario, it may be impossible for the controller to find routing paths with sufficient available capacity for all flows at all times. Mathematically, the optimal solution to the MLU minimization problem will be $\theta > 1$ which means that the capacity of at least one link will be exceeded. Even when a solution $\theta < 1$ can be found it does not mean that congestion can be always prevented. Short-scale traffic bursts can induce a rapid build-up of queued packets at a bottleneck link, leading to dropped packets and a reduced transmission rate when the queue is exhausted. These short-term traffic dynamics cannot be adequately addressed by the centralized controller which typically selects routing paths for flows in a longer time scale.

To address the above challenges, routing path control is usually complemented with scheduling and traffic control mechanisms. The purpose of these mechanisms is to exercise control over the time and volume of packets that are transmitted throughout the network rather than which paths are used for routing these packets. These decisions can change quickly to adapt to network dynamics, faster than the time it would take to replace the flow rules for changing the routing paths. The scheduling and traffic control decisions can be made by both the switches and the endpoints (known as hosts) in the network. In fact, the very first scheduling and traffic control ideas go back to the beginning of the Internet and the invention of the TCP protocol which makes flow control decisions at the network endpoints. With SDN, particularly in data center network, these decisions can be made by the centralized controller and realized as rules in the switches similar to the routing decisions.

In this section, we discuss the role of SDN and review the previous work in the area of scheduling and traffic control. We categorize the previous work into two main groups: (i) flow control methods in datacenters and (ii) Backpressure-based methods. On the one hand, scheduling flows among servers in datacenter networks has been shown to be an effective approach to achieve fair and consistent flow throughputs. On the other hand, Backpressure is a scheduling algorithm that has received significant attention from the networking research community and applied to various domains such as sensor networks, mobile ad hoc networks, and heterogeneous networks with wireless and wireline components.

3.3.1   FLOW CONTROL IN DATACENTERS

TCP Protocol

Transmission Control Protocol (TCP), the widely used Internet transport protocol, uses the idea of a congestion window to indirectly affect the packet scheduling decisions. The size of this window can be dynamically adjusted based on feedback (acknowledgment) information sent by the receiver to the sender of each packet. This allows adapting the rate of packet transmission between each pair of sender-receiver hosts—decrease the window in times when the network is becoming congested so as to slow down further transmissions, and gradually increase it again when congestion is abated.

The Role of SDN

SDN can be used to improve the scheduling decisions in various ways. First, the SDN controller can be employed to centrally collect information about the state of the network and subsequently compute and distribute congestion control parameters such as the initial and maximum window size as well as re-transmission timers used by the TCP protocol. This has been shown to be quite effective in improving the overall network performance and resource utilization by the TCP protocol [145]. Second, the SDN controller can explicitly determine the scheduling decisions, when exactly each host's packets should be transmitted. Such decision-making logic can be

Table 3.2: State-of-the-art works on scheduling in datacenters

Ref.	Acronym	Granularity of Control	Scheduling	Routing	Method
[146]	Fastpass	Packet-level	✓	✓	Matching and Edge coloring
[147]	Hedera	Flow-level	✓	✓	Global first fit, Simulated annealing
[148]	TDMA	Flow-level	✓	✗	Weighted round robin
[149]	Flowtune	Flowlet-level	✓	✗	Network utility maximization
[150]	TAPS	Flow-level	✓	✓	Deadline-aware algorithm

realized not only at the hosts but also at the switches in a data center network. This allows us to exercise tighter control for transmission scheduling for each link and each flow in the network.

The scheduling problem (and the role of SDN therein) has been extensively studied in the recent past. Table 3.2 summarizes some of the main work in this area. Most of these works target data center networks since these networks exhibit a number of unique challenges that make scheduling decisions more critical. Specifically, in data centers there is a rich mix of workloads often with different priorities and deadlines while congestion can easily happen with queuing often dominating end-to-end latencies. This is different from WANs where propagation latency determined by the selection of the routing paths is non-negligible and a critical factor to taken into account. Some of the work considers only the scheduling of transmissions but not the routing decisions (the routing paths are assumed to be fixed and given). Another difference is that some of the work makes transmission decisions on a per packet basis while others do so on a per flow basis (e.g., for each pair of sender-receiver hosts or other features that define a flow). The granularity of control is important and determines the effectiveness of the scheduling decisions in each scenario.

FASTPASS Scheduler

We next provide a detailed discussion on the work in [146], named FASTPASS, one of the influential and frequently cited papers. This paper proposes specific algorithms for the scheduling and routing problems. It considers a data center network like the one depicted in Figure 3.7. This in fact shows the structure of a typical data center network where groups of machines, or racks, at the bottom layer (where the endpoints/hosts reside) are connected to a Top of Rack (ToR) switch which provides non-blocking connectivity to them. ToR switches are then connected via a large interconnection network allowing machines to communicate across racks. Although

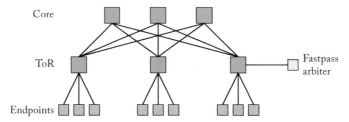

Figure 3.7: Fastpass arbiter in a two-tier data center network topology.

SDN is not explicitly mentioned, the paper proposes that each packet's timing and routing path are controlled by a logically centralized arbiter, which can be well implemented by an SDN controller. By exercising such tight control, we can enforce the flow rates to match the available network capacity over the time scale of individual packet transmission times, unlike the distributed congestion control mechanisms that take longer times to converge. This way, not only persistent congestion will be eliminated but also queues where packets are accumulated before transmission over the links will never vary in size, latencies will be stable, and packets will never be dropped due to queue overflow.

Fastpass includes algorithms for scheduling and routing of packets. For scheduling, it proposes a fast and scalable timeslot allocation algorithm to determine when each endpoint's packets should be sent. For routing, it proposes a path selection algorithm to select a path for each packet that is scheduled for transmission. Note that these decisions are made separately. First the scheduling is made subject to input and output bandwidth constraints at the endpoints and, given that, the routing inside the network follows. In order for this approach to work, it assumes that there is always available bandwidth capacity inside the network to route any traffic scheduled without queueing. Therefore, the bottleneck is at the endpoints, a property known as rearrangeably non blocking (RNB).

Some classic optimization methods from graph theory are proposed to address the scheduling and routing problems. Namely, the scheduling (timeslot allocation) can be solved as a matching problem while the routing (path selection) as an edge coloring problem. The two algorithms are illustrated in Figure 3.8 and discussed below.

Figure 3.8a shows the matching problem. There exist four endpoints and five pairs of them (arrows) have pending demands for packet transmission. The timeslot allocator orders the demands of the different source-destination endpoints by the last timeslot allocated to each pair. On the right is the state used to track bandwidth constraints: one bit for each source and for each destination. The first two demands can be allocated because both the source and destination are available (illustrated as white squares), but the third demand cannot be allocated because destination 3 has already been allocated. The remaining two demands can be allocated, yielding a maximal matching. From the literature on input-queued switches, it is well known that any maximal matching provides 50% throughput guarantees.

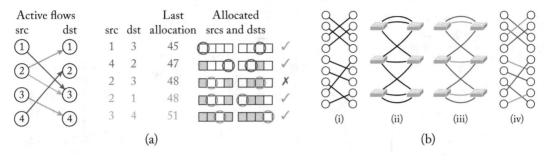

(a)   (b)

Figure 3.8: (a) Timeslot allocation and (b) path selection; input matching, ToR graph, edge-colored ToR graph, and edge-colored matching (from left to right).

In Figure 3.8b, we depict the path selection process. The goal of path selection is to assign packets to paths such that no link is assigned multiple packets in a single timeslot. This property guarantees that there will be no queueing within the network. In a network with two tiers, ToR and core, with each ToR connected directly to a subset of core switches (Figure 3.7) each path between two ToRs can be uniquely specified by a core switch. Thus, path selection entails assigning a core switch to each packet such that no two packets (all of which are to be sent in the same timeslot) with the same source ToR or destination ToR are assigned the same core switch. This assignment can be performed with graph edge coloring. The edge-coloring algorithm takes as input a bipartite graph and assigns a color to each edge such that no two edges of the same color are incident on the same vertex. The data center network can be modeled as a bipartite graph where the vertices are ToR switches, edges are the allocated packets, and colors represent core switches/paths. The edge-coloring of this graph provides an assignment of packets to paths such that no link is assigned multiple packets. In Figure 3.8b, the matching of packets to be transmitted in a given timeslot (a) is transformed into a bipartite multi graph of ToRs (b), where the source and destination ToRs of every packet are connected. Edge-coloring colors each edge ensuring that no two edges of the same color are incident on the same ToR (c). The assignment guarantees that at most one packet occupies the ingress, and one occupies the egress, of each port (d). A network with n racks and d nodes per rack can be edge-colored in $O(nd \log d)$ time.

3.3.2   BACKPRESSURE-BASED METHODS

Backpressure-based methods schedule packet transmissions across links in a network so as to balance the load in virtual queues at the nodes where packets are accumulated before being transmitted. The decisions are driven by the congestion status in the network adding pressure to the packets to flow from heavier to less loaded queues. While distributed implementations exist, most Backpressure methods are centralized and therefore can largely benefit by the capabilities of SDN. In the following, we discuss the original as well as various extensions of the Backpressure algorithm.

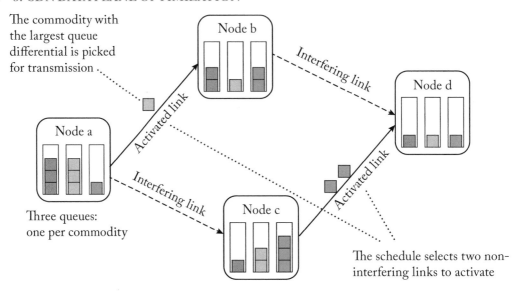

Figure 3.9: An example illustrating the virtual queues in Backpressure algorithm.

Original Algorithm

The original Backpressure algorithm was proposed in [153] over 30 years ago and is one of the most influential works in networking literature with over 3,000 citations by the year 2021. The algorithm works in a time slot-based manner. At each time slot, it activates a set of non-interfering links in the network, which leads to the maximum sum of link weights multiplying the respective link rates. The definition of interfering links depends on the interference model. For example, under the simple 1-hop interference model (also known as the protocol model), two links having a common endpoint node interfere with each other and therefore they cannot successfully transmit packets at the same time. The key idea of the Backpressure algorithm lies in the way the link weights are chosen. Specifically, the weight of a link is defined as the differential of the queue backlogs between the two endpoints of that link. Thus, the weight will be large for a link for which many packets have been accumulated at its first endpoint (sender) and wait for transmission to the second endpoint (receiver). The rates represent the bandwidth capacities of the links (bps). Therefore, the algorithm "adds pressure" to packets to be transmitted to from heavy to less loaded queues that are connected via high rate links.

The Backpressure algorithm can be applied to multi-commodity networks where packets of different commodities need to be transmitted and may have different sources and destinations as well as different quality of service requirements. In this case, the packets of each commodity are accumulated at separate queues and the link weight is defined as the maximum of the queue differential over all commodities. The operation of the Backpressure algorithm is illustrated in Figure 3.9. Formally, at the beginning of time slot t, the weight of a link (a, b) is defined as:

$$W_{ab}(t) = \max_{c} \left(Q_a^c(t) - Q_b^c(t) \right), \tag{3.28}$$

where $Q_a^c(t)$ represents the queue backlog of commodity c on node a at time t. The backpressure algorithm will activate for transmission the links in the schedule $\pi(t)$ derived from the following optimization problem:

$$\pi(t) = \underset{\pi \in \Pi}{\text{argmax}} \sum_{(a,b) \in \pi} W_{ab}(t) r_{ab}(t), \tag{3.29}$$

where Π represents the set of all possible feasible (non-interfering) schedules, and $r_{ab}(t)$ denotes the link rate. The work in [153] proved that the schedule computed based on the above two equations is throughput optimal in the sense that it can support any arrival rates that are supportable by any other scheduling policies. The proof is established using Lyapunov optimization theory.

Finding the optimal schedule in Equation (3.29) requires global network state information about the link rates ($r_{ab}(t)$ values) and queue backlogs ($W_{ab}(t)$ weights). The SDN controller can centrally collect this information and solve the problem. Solving this problem, however, has high computational complexity; cubic to the number of nodes in the network under the one-hop interference model, and even NP-hard for more general interference models. To compute a solution in reasonable time, heuristic algorithms have been proposed. One of them is the Greedy Maximal Scheduling (GMS) that greedily activates non-interfering links with high weights which has been shown to achieve at least half of the optimal throughput performance.

We emphasize that because of the communication delay between the SDN controller and the data plane nodes, in practice the controller can only collect outdated information about the network state. That is, at each time slot t the controller will be aware of the outdated values $r_{ab}(t - \Delta)$ and $W_{ab}(t - \Delta)$ rather than the current ones, where Δ is the number of slots that information of link (a, b) took to reach the controller. This delay may be different for different links and will in general increase with the distance (e.g., number of hops) to the controller. The outdated state may affect the throughput optimality of the Backpressure algorithm. This issue was explored in [89] where a method was proposed to place the controller in a location in the network so that to mitigate the effects of the delayed collected state information, as we discussed in Section 2.1.6.

It is also worth clarifying that the scheduling decisions made by the Backpressure algorithm determine also the routing paths. Indeed, as queues are scheduled for transmission over links, the packets inside them are routed to the queues in the next hop nodes, hop by hop until they reach the destination node. Such joint scheduling and routing methods, also known as cross-layer, advanced the design of network protocols over recent years.

Algorithm Extensions
Various extensions of the Backpressure algorithm have been proposed. A survey of these works is available in [154]. An area of research focused on distributed algorithm implementations where

each node in the network makes its own scheduling decisions based on local state information, thereby improving scalability. A lot of research activity has been also devoted to achieve desirable delay performance. Since packets are routed in the network according to congestion information only, they may take unnecessary long paths and/or even loops, which results in large packet delivery delay. To mitigate this issue, variants of Backpressure algorithm encourage packets to travel along shortest paths, e.g., by imposing a bias on these paths. Another extension was presented in [155] to handle a mix of unicast, broadcast, multicast, and anycast traffic. A novel aspect of this algorithm, called Universal Max-Weight (UMW), is that it uses a source routing logic to prescribe an admissible route to each incoming packet immediately upon its arrival rather than in a hop by hop manner. This way, packet cycling is guaranteed to never happen while throughput optimality is achieved. Finally, the scenario were the Backpressure algorithm can control the decisions of only a part of the nodes in the network, whereas the rest nodes are controlled by an unknown protocol, was considered in [156]. An online algorithm that attempts to learn the decision-making logic of the uncontrollable nodes was proposed and the three-way tradeoff between throughput, delay and convergence rate of the learning process was explored.

3.4   NETWORK SLICING

In addition to traffic engineering and scheduling mechanisms, SDN can be used to support network slicing. Network slicing is a new network paradigm where multiple logical (virtual) networks are created over a common network infrastructure each of them devoted to a different service and customized to its specific needs. These logical networks are referred to as network slices. The concept of separated virtual networks deployed over a network is not a new one; it is known since the inception of virtual private networks (VPNs) traced back in the 1990s. However, there are specificities that make network slicing a novel concept. Namely, a network slice needs to create the illusion to the slice tenant of operating its own dedicated physical network isolated from other slices. In addition, the slices can be self-contained with independent control and management as well as be flexible enough to dynamically accommodate diverse requirements of services. The requirements of the services can be diverse and involve different types of resources such as computation, storage, and communication.

Within a slice, virtual nodes are connected to each other through virtual links. A virtual link can be realized in the physical network as a multihop path with reserved bandwidth on all physical links constituting the path. A virtual node implements a specific network function (NF) on a physical node such as firewall, deep packet inspector or video encoder. SDN provides a high degree of flexibility to realize the virtual links. An SDN controller can remotely configure the physical network to reserve on demand the bandwidth needed by the virtual links in the slice and install the corresponding flow rules at the SDN switches to enforce the routing of traffic over the corresponding physical paths. Complementary to SDN, another technology referred as Network Function Virtualization (NFV) can be used to flexibly realize the virtual nodes. With NFV, the virtual nodes can be realized as virtual functions which can be run on general-purpose hardware

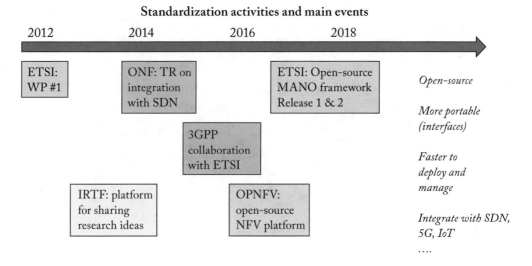

Figure 3.10: Standardization activities and main events of NFV.

and be instantiated or migrated from one physical node to another with relative ease depending on the actual slice requirements. This is radically different from the traditional physical functions that are tied to specific customized hardware, also known as middleboxes, and are less flexible to dynamically deploy on-demand on a slice-specific basis.

While the integration of SDN and NFV allows us to realize network slices in a more flexible manner, the question of how to efficiently allocate, manage, and control the slice resources (computation, storage, and communication) in real time raises new challenges. In the sequel we focus on the algorithmic challenges that emerge in efficient network slicing, necessitating novel modeling and optimization methods. Before that we present an overview of the NFV technology that is needed to realize the virtual functions in the network.

3.4.1   NETWORK FUNCTION VIRTUALIZATION

History

NFV is a technology that decouples network functions from the underlying dedicated hardware and realizes them in the form of software, which are referred to as Virtual Network Functions (VNFs). This technology started from over 20 of the world's largest network operators back in 2012. The main idea was that network operations can be simplified and costs can be reduced if all network functions become virtualized as software appliances instead of the traditional middleboxes that require dedicated hardware. The Industry Specification Group (ISG) was formed within the European Telecommunications Standards Institute (ETSI) to define this technology. This group has grown to a large community with 300+ members all over the world by 2021.

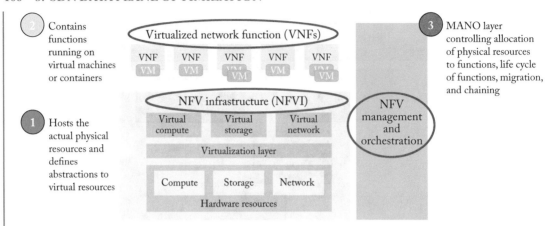

Figure 3.11: ETSI NFV reference architecture.

Since its inception in 2012, NFV has evolved substantially. Figure 3.10 shows a non-exhaustive list of standardization activities and main events that contributed to NFV evolution. ETSI released several white papers that cover many aspects of NFV including management and orchestration of the virtual functions, architectural framework, and service quality metrics. In parallel, the Open Networking Foundation (ONF), which primarily focuses on innovating network operation through SDN, published a technical report on how SDN can be integrated with NFV so that one technology can be used to enable the other. The following years various other organizations like 3GPP, IRTF, and OPNFV contributed toward developing an open-source NFV management and orchestration software framework, interoperable and portable NFV solutions through appropriate interfaces and making them more suitable for 5G networks. Furthermore, decreasing the overheads of the virtual functions and allowing them to run in lightweight devices increased suitability for resource-constrained network environments like Internet of Things (IoT).

Architecture

The NFV architecture is depicted in Figure 3.11. It contains three layers. First, the Network Function Virtualization (NFVI) layer corresponds to the data plane of the network that forwards data and provides resources (computation, storage, and network) for running network services. Second, the VNF layer corresponds to the application plane, which hosts various kinds of VNFs that can be regarded as applications. The VNFs require a mechanism to allocate resources to them according to their application needs. This is handled by the third layer referred to as Management and Orchestration (MANO). This layer corresponds to the control plane and is responsible for building the connections among various VNFs in the VNF layer and orchestrating resources in NFVI.

SDN-NFV Integration

Although SDN and NFV were initially developed as independent networking paradigms, the evolution of both technologies has shown strong synergy between them. Both technologies share a common goal of making networks more programmable and flexible but from different view points. SDN decouples data and control planes while NFV decouples service functions and network infrastructures. Therefore, SDN and NFV are complementary to each other. By integrating the two technologies benefits can be realized for both of them. For example, NFV can be used to slice the data plane infrastructure. Virtualization can be also used so that different tenants take slices of resources of a shared SDN controller to run their own control applications separately.

3.4.2   ALGORITHMIC CHALLENGES

The network slicing paradigm makes network management flexible and introduces new network optimization problems. In addition to the transport of traffic we discussed in the previous sections, optimized placement of virtual functions and the allocation of resources to run these functions on the available network nodes are required. Besides, to facilitate service delivery, multiple of these functions may be chained in a predefined order and traffic should be steered through them giving rise to the concept of *service function chaining*. The SDN controller can be used to centrally compute the solution to such optimization problems. In the sequel, we discuss the algorithmic challenges posed by these kind of problems.

Service Function Chaining

The Service Function Chaining (SFC) problem is central to the network slicing paradigm. Our ability to effectively deploy SFCs can empower network slicing. In other words, SFC is a key enabler for network slicing. In a nutshell, a SFC describes a sequence of service functions that a traffic flow needs to pass through in a particular order so as to get a complete end-to-end service. For example, as depicted in Figure 3.12a, a service chain could define that a flow is first routed from its source (a handheld connected to a cellular base station) to a firewall for security, next through a deep packet inspection (DPI) function and a load balancer for traffic optimization, and only then delivered to its destination (a web server). SDN facilitates the dynamic on-demand realization of such chains by providing flexible and efficient forwarding of traffic among the network nodes hosting these functions.

The above scenarios give rise to a new problem that is more complex than the traffic engineering and scheduling discussed in the previous sections. Specifically, the network operator needs to decide not only the routing of traffic flows through the appropriate function instances but also the placement of functions in the network along with the chaining of them. This problem is further complicated by the multidimensional resource requirements of the service functions and the way they are chained together. Consider for example an Augmented Reality type of service, as in Figure 3.12b, where flows from two video cameras on the left of the figure go

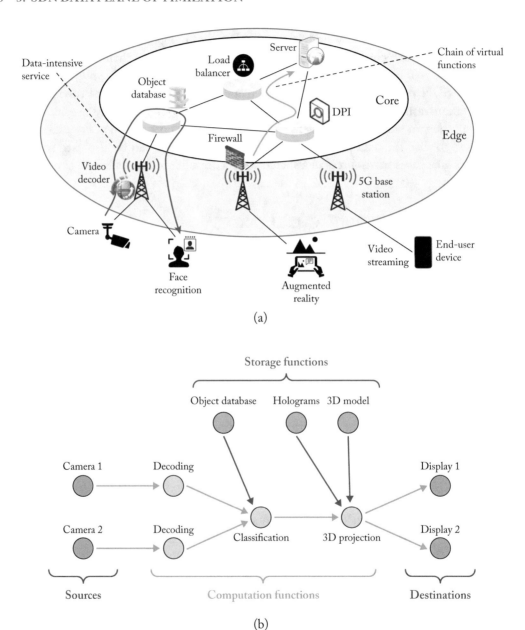

(a)

(b)

Figure 3.12: (a) An illustrative network where traffic is steered through chains of service functions instantiated at core and edge nodes. (b) An example of an augmented reality type of service chain.

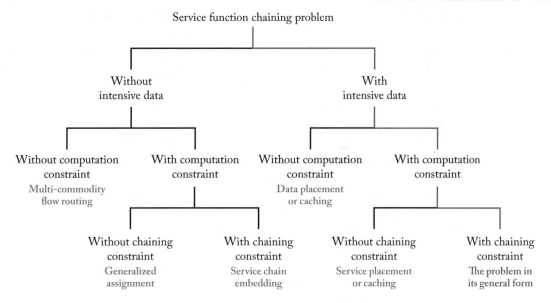

Figure 3.13: Classification of the related work for the service function chaining problem.

through chains of functions for decoding, classification and 3D projection so as to adapt the rates of the video streams, detect objects in them and augment holograms to them, before finally delivering the augmented streams for display at two consumer devices on the right in the figure. Note how these operations require computation and communication resources for executing processing tasks and delivering associated outputs to the end users. They also require non-trivial amounts of data that should be stored in advance in the network and be available for the service functions to access (e.g., object databases for classification, holograms and 3D models for 3D projection) in an on demand manner.

Classification of State-of-the-Art Work

A lot of work in the literature studied the service function chaining problem with most of them making various simplifying assumptions to facilitate its solution. A categorization of these works is in Figure 3.13.

A first important assumption made by many of these works is that the resources required by the service functions are dedicated. This means that each function takes up its own portion of resources for exclusive use and the total resource consumption at a node or a network link is the sum of resource allocations. Computation and communication resources are typically allocated in this manner. However, the storage resource has a shareable (non-dedicated) nature. Namely, a stored data object can be shared to satisfy multiple services with the same or overlapping data requirements (e.g., the same object database or 3D model in the service example in Figure 3.12b).

Such sharing is important as we do not need to pay for resources multiple times; one time is enough. This is an important difference from a modeling and optimization point of view and requires new problem formulations. The existing works that treat all resources as dedicated fail to utilize the opportunities for storage resource sharing. Such opportunities exist more often for modern data-intensive services like Face Recognition and Augmented Reality that require storage and access to large-volumes of data. However, other traditional middlebox-like services such as firewalls, network address translators, load balancers, and video encoders do not require a lot of data and therefore the assumption of dedicated resources is realistic.

In the extreme case that services need negligible computation (or when computation capacity of the nodes is not the bottleneck), then the only requirement is to route traffic flows over the capacitated links among functions in a network. This can be cast as a variant of the *multi-commodity flow routing problem* [158]. For the more practical case that functions require to execute some computation task, the work in [159] formulated the computation and communication allocation problem as a *generalized assignment problem* for which approximation algorithms are known. Subsequent works in [160] and [161] proposed integer linear programming formulations for the objectives of minimizing cost or maximizing accepted requests, and solved the problem using bargaining and conformal mapping theory, respectively.

The problem is further complicated, however, if one considers order constraints in the execution of the functions known as chaining constraints. Chained service functions can be modeled as service graphs (like the one in Figure 3.12a) to be *embedded* into a physical network and various algorithms have been proposed for this purpose using techniques such as randomized rounding [162], relaxation and successive convex approximation [163], and shortest path routing on a multi-layer graph [164]. Alternative modeling and solution approaches are also known based on mixed integer linear programming [165], [166], column generation [167], sampling-based Markov approximation [168], and queuing models [169]. However, the above works did not consider functions with intensive data requirements or treated data storage as another dedicated non-shareable resource to allocate, just like computation and communication.

As stated previously, many of the newer services require access to non-trivial amounts of data (e.g., data base of images, visual models, etc.) which should be stored in advance at the nodes and be available for the functions to share in an on demand manner. This can be formulated as a *data placement problem* [170] also referred to as *caching problem* [171], [172] and is among the most well-investigated problems in networking literature. However, the data placement/caching problem is much simpler than the service chaining problem since it does not deal with the allocation of computation resources, the execution of functions and the accompanying chaining constraints. Still, this problem is NP-Hard due to the combinatorial nature of the data placement decisions.

Recently, there has been some work that generalizes the data placement problem by adding computation allocation variables and computation constraints to satisfy requests for data-intensive services [173], [174], [175]. This generalized problem is referred to as the *service place-*

ment problem and several approximation algorithms for jointly allocating storage, computation, and communication resources have been proposed. While these works considered the shareable nature of the stored data, they assumed that each service request is for a single function rather than a chain of functions, and that data objects must be colocated with the services using them at the same cloud location. Besides, these works focused on special network topologies of two tiers. In the sequel, we present a general model and problem formulation, originally presented in [176], that captures all the above aspects: data-intensive nature of services; shareable nature of data storage, chaining, computation, and bandwidth capacity constraints; and arbitrary network topologies.

General Problem

We begin by describing the network and service models. Next, we present the mathematical formulation, solution algorithm, and evaluation results.

Network Model

We model the network as a directed graph $G = (V, \mathcal{E})$ with V vertices and $\mathcal{E}$ edges representing the set of network nodes and links, respectively. A node in V may represent an end-user device, an access point, or a cloud node inside the core or edge network. Each node $u \in V$ is characterized by its computation resources (e.g., CPUs or GPUs) and storage resources (e.g., hard disk or flash memory). We denote by C_u and R_u the computation and storage capacities at node $u \in V$, respectively. The cost of allocating one unit of processing and storage resource at node u is c_u and r_u, respectively. Nodes are interconnected via wireless or wireline links, each characterized by its transmission capacity and unit cost. Link (u, v) has transmission capacity B_{uv} and the cost per bandwidth resource unit is b_{uv}.

Service Model

A generic service $\phi \in \Phi$ can be described by a directed acyclic graph (DAG) $G^\phi = (V^\phi, \mathcal{E}^\phi)$ where vertices represent service functions and edges represent streams of information exiting one function and entering another. An example of a service graph is shown in Figure 3.12b. There are four types of vertices in the service graph G^ϕ.

1. *Source or production functions* ($V^{\phi,s}$ set) are vertices with no incoming edges in the service graph. They represent sensors that generate real-time data streams (e.g., video cameras, IoT sensors) associated with specific locations in the physical network (e.g., user devices, nearby access points).

2. *Storage functions* ($V^{\phi,r}$ set) are also vertices with no incoming edges in the service graph. They represent the storage/caching of pre-generated static data (e.g., pre-coded video, database). There is one storage function for each computation function that requires access to a given data object. Hence, if multiple computation functions require access to the same data object, we still create multiple storage functions.

3. *Computation functions* ($\mathcal{V}^{\phi,c}$ set) are internal vertices of the service graph. They take as input source, storage, or other computation functions, and pass their output to other computation or destination functions.

4. *Destination or consumption functions* ($\mathcal{V}^{\phi,d}$ set) are vertices with no outgoing edges in the service graph. They represent end devices (e.g., video displays), consuming output information streams and are associated with specific locations in the physical network, just like the source functions.

We define the storage and computation requirements of the service functions (vertices of the service graph) as follows. A storage function $i \in \mathcal{V}^{\phi,r}$ has zero computation requirement but requires storing a data object (e.g., video, object database, holograms for an AR service, etc.) which consumes storage resources. We denote by $o(i)$ and $r^{o(i)} > 0$ the respective data object and the required storage resource, where the latter clearly depends on the size of the data object. A computation function $i \in \mathcal{V}^{\phi,c}$ has zero storage requirement but requires computation resources $c^i > 0$ for task execution. The rest of the vertices (source and destination functions) have zero storage and computation requirements.

Next, we define the bandwidth requirements of the service streams (edges of the service graph). We denote by b^{ij} the bandwidth requirement of stream $(i, j) \in \mathcal{E}^{\phi}$. For streams generated by a source or computation function i, b^{ij} depend on the volume of generated data (e.g., video bitrate). For streams coming out of a storage function i, b^{ij} depend on the size of the stored data and the rate at which function j accesses the data.

To facilitate presentation, we denote by $\mathcal{G}^{\Phi} = (\mathcal{V}^{\Phi}, \mathcal{E}^{\Phi})$ the union of all service graphs, where $\mathcal{V}^{\Phi} = \cup_{\phi} \mathcal{V}^{\phi}$, and $\mathcal{E}^{\Phi} = \cup_{\phi} \mathcal{E}^{\phi}$. Similarly, $\mathcal{V}^{\Phi,s}$, $\mathcal{V}^{\Phi,r}$, $\mathcal{V}^{\Phi,c}$, and $\mathcal{V}^{\Phi,d}$ denote the unions of all source, storage, computation, and destination functions for all service graphs. We also denote by $\mathcal{S}(u)$ and $\mathcal{D}(u)$ the sets of all source and destination functions associated with a particular node u. Finally, we denote by $\mathcal{O} = \cup_i o(i)$ the union of all data objects.

Mathematical Formulation

The network needs to decide how to embed (or map) the vertices and edges of the collection of service graphs to the nodes and links of the physical substrate network. To model the embedding decisions, we introduce two sets of optimization variables: (i) $x_u^i \in \{0, 1\}$ which indicates whether vertex i is mapped to physical node u ($x_u^i = 1$) or not ($x_u^i = 0$), and (ii) $y_{uv}^{ij} \in \{0, 1\}$ which indicates whether edge (i, j) is mapped to a path that uses physical link (u, v) ($y_{uv}^{ij} = 1$) or not ($y_{uv}^{ij} = 0$).

Differently from traditional service embedding problem formulations (e.g., [162]), in this case, the network needs to also decide the data placement decisions: $z_u^o \in \{0, 1\}$ which indicates whether data object o is stored at physical node u ($z_u^o = 1$) or not ($z_u^o = 0$). The respective vectors of optimization variables are the following:

$$x = (x_u^i \in \{0, 1\} : u \in \mathcal{V}, i \in \mathcal{V}^\Phi) \tag{3.30}$$

$$y = (y_{uv}^{ij} \in \{0, 1\} : (u, v) \in \mathcal{E}, (i, j) \in \mathcal{E}^\Phi) \tag{3.31}$$

$$z = (z_u^o \in \{0, 1\} : u \in \mathcal{V}, o \in \mathcal{O}). \tag{3.32}$$

The above decisions need to satisfy several constraints. First, we need to make sure that each source and destination function is mapped to its associated node in the physical network:

$$x_u^i = 1, \ \forall u \in \mathcal{V}, i \in \mathcal{S}(u) \cup \mathcal{D}(u). \tag{3.33}$$

Second, we need to map each storage and computation function to exactly one node in the physical network:

$$\sum_{u \in \mathcal{V}} x_u^i = 1, \ \forall i \in \mathcal{V}^{\Phi,r} \cup \mathcal{V}^{\Phi,c}. \tag{3.34}$$

Third, we need to ensure that each edge $(i, j) \in \mathcal{E}^\Phi$ is mapped to a path in the physical network that starts at the location of function i and ends at the location of function j, hence respecting service chaining requirements:

$$\sum_{(u,v) \in \mathcal{E}} y_{uv}^{ij} - \sum_{(v,u) \in \mathcal{E}} y_{vu}^{ij} = x_u^i - x_u^j, \ \forall u \in \mathcal{V}, (i, j) \in \mathcal{E}^\Phi. \tag{3.35}$$

Fourth, the total computation load of services running on physical node u must not exceed its computation capacity:

$$\sum_{i \in \mathcal{V}^{\Phi,c}} x_u^i c^i \leq x_u \leq C_u, \ \forall u \in \mathcal{V}, \tag{3.36}$$

where x_u is an auxiliary variable whose value is equal to the computation load of node u in the optimal solution of the problem. Fifth, the total bandwidth load of services transported over physical link (u, v) must not exceed its bandwidth capacity:

$$\sum_{i,j \in \mathcal{E}^\Phi} y_{uv}^{ij} b^{ij} \leq y_{uv} \leq B_{uv}, \ \forall (u, v) \in \mathcal{E}, \tag{3.37}$$

where y_{uv} is an auxiliary variable whose value is equal to the bandwidth load of link (u, v) in the optimal solution.

To satisfy the data requirements of services, we also need to satisfy the following two sets of constraints:

$$x_u^i \leq z_u^{o(i)}, \ \forall u \in \mathcal{V}, i \in \mathcal{V}^{\Phi,r} \tag{3.38}$$

$$\sum_{o \in \mathcal{O}} z_u^o r^o \leq z_u \leq R_u, \ \forall u \in \mathcal{V}. \tag{3.39}$$

The first of the two constraints ensures that a storage function i can be mapped to a node u ($x_u^i = 1$) only if the latter has stored the required data object $o(i)$. The second set of constraints ensures that the total storage capacity of a node u is not exceeded, where z_u is an auxiliary variable whose value is equal to the storage load of node u in the optimal solution.

The goal of the network operator is to make the embedding (x, y) and data placement (z) decisions that minimize the total resource consumption costs:

$$\min_{x,y,z} \quad \sum_{u\in\mathcal{V}}(x_u c_u + z_u r_u) + \sum_{(u,v)\in\mathcal{E}} y_{uv} b_{uv} \tag{3.40}$$

$$s.t. \qquad \text{constraints: } (3.30) - (3.39).$$

Approximate Solution and Evaluation

Since the problem above is NP-Hard, the use of approximation algorithms is justified. A popular method to derive such an approximation algorithm is by solving the linear relaxation of the problem and then rounding the optimal fractional solution found to an integer solution that is feasible for the original problem. The main challenge in designing and analyzing such algorithms lies in the way the rounding decisions are made. Using randomization can help in analyzing the performance of rounding decisions compared to a more rigid deterministic rounding approach (e.g., round the value of each variable to 1 if the fractional value is above a threshold). The rationale is that it may be possible for the outcome of random decisions to be bounded as the scale of the problem increases by using well-known stochastic inequalities and theorems. This provides an additional tool toward arguing how much the rounded solution value worsens compared to the optimal fractional value, and also whether and the extent to which the problem's constraints are violated.

Yet, the design of such a randomized rounding algorithm is not straightforward, especially when the problem contains multiple constraints that mix variables in a complex manner. In [176], such an algorithm was designed by extracting multiple embeddings from the fractional solution. In an iterative manner, an embedding that corresponds to a fractional solution with strictly positive variable values is extracted, and the fractional solution values are reduced accordingly until no more embeddings can be found. In the end an embedding is chosen randomly. Then, analysis can be done using Hoeffding's Lemma and the union bound theorem.

The performance evaluation of the randomized rounding algorithm in [176] is discussed below. The setup is depicted in Figure 3.14a. Services can be hosted in different tiers in a network: a head office (HO) node at the higher tier that represents a centralized data center, intermediate office (IO) and end-office (EO) nodes, and base station (BS) nodes associated with mobile end-users at the bottom tier. The service chains follow the structure shown on the right of the figure. Each chain contains a source and a destination vertex associated with specific locations in the network as well as storage and computation functions, the location of which can be optimized. This structure can capture various data-intensive services such as Augmented Reality (AR).

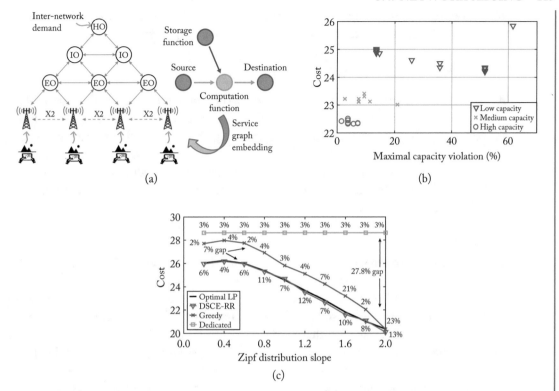

Figure 3.14: (a) Evaluation setup of 10 nodes and 36 directed links (left) and service graph of 4 vertices and 3 edges (right). (b) Scatter plot of cost vs. maximal capacity violation across all the embeddings extracted by DSCE-RR algorithm. (c) Different slopes of the service demand distribution with labels representing the respective maximal capacity violation factors.

The results for embedding 100 services are shown in Figure 3.14b for 3 scenarios of different resource capacities; low, medium, and high. We note that multiple embeddings are extracted by the algorithm in each scenario, each embedding corresponding to a different pair of cost and capacity violation. There is a clear trend that the capacity violation factor decreases (and the solution approaches feasible) as capacities increase. Besides, Figure 3.14c highlights the impact of the slope of the data demand distribution, i.e., the more skewed the distribution is the more likely that many service requests require the same data object and hence the more the benefit of sharing that data object to reduce resource consumption. Here, in addition to the performance of the randomized rounding algorithm, referred as DSCE-RR, two more baseline algorithms are evaluated as well as the optimal solution to the linear relaxation of the problem that provides a lower bound to the optimal integer solution. The two baselines, greedy and dedicated, allocate the most popular functions to the base stations—close to the demand—and ignore the data sharing opportunities among functions, respectively. The fact that DSCE-RR performs better than the two baselines verifies that the solution is not straightforward.

3.5   NETWORK SECURITY

Security is always an important aspect that needs to be considered in networks and SDN can play an important role in this area. Traditional methods to secure a network involve the deployment of physical and application layer protection schemes, e.g., by strong authentication and encryption during data transmission. Complementary to the above, network layer security methods are additionally used to monitor the events in a network system and identify possible attacks. Such methods are usually referred to as Intrusion Detection Systems (IDS). In this section, we discuss an SDN-based approach for IDS.

Types of IDS

Broadly there are two types of IDSs—signature-based IDS and anomaly-based IDS. Signature-based IDSs inspect the payload of the packet and compare against signatures of known attacks to identify potential malicious activities. On the other hand, anomaly-based IDSs use a statistical method that creates a model based on collected data of the behavior of legitimate and malicious users and compares packets with that model to identify if there is significant deviation from it. The advantages of an anomaly-based IDS compared to signature-based IDS include its ability to detect new attacks (without a known signature) as well as that it requires inspection of only flow-level information such as packet header rather than the whole packet payload.

SDN for Anomaly-Based IDS

The capabilities of SDN facilitate the deployment of anomaly-based IDSs to enhance network security. First, the logically centralized control and global network view features of SDN simplify the collection and analysis of network traffic to build the corresponding statistical model. Second, the programmability of SDN makes it easy to react to network events immediately when they are detected by dynamically updating the flow rules in the data plane switches so as to match and block malicious traffic according to the statistical model. The controller is also able to modify flow tables in switches in advance to prevent network attacks and limit the communication between the control plane and the data plane. Note that the above process mimics the way that traditional middlebox-like firewall services work but this time with a new flexible and dynamically reconfigurable aspect due to the new capabilities that SDN brings.

The Role of Machine Learning

Machine Learning (ML) can be used to generate the statistical model and it has been demonstrated as a promising method for identifying attacks even with unknown or encrypted traffic flows. In fact, the intrusion detection problem can be considered as a classification task. Thus, supervised learning algorithms can be applied to address this problem. In the simplest case, the training features used by the ML algorithm can be the specific header fields of the packet. For example, [177] proposes an HMM-based Network Intrusion Detection System (NIDS) with five training features (the length of the packet, source port, destination port, source IP address, and destination IP address). In [178], deep NN model is used in SDN to detect intrusion activities, by classifying traffic flows into normal and anomaly classes. The deep NN model has an input layer, three hidden layers, and an output layer. [179] uses deep recurrent NN in an anomaly-based IDS and proposes a Gated Recurrent Unit Recurrent Neural Network (GRU-RNN) algorithm to detect intrusion. To speed up intrusion detection and reduce the computational cost, only six flow features such as duration and protocol type are used to train the GRU-RNN algorithm. [180] applies learning using as training features the packet headers of the 6LoWPAN protocol.

While the above methods are effective, heterogeneous network protocols may have distinct packet header structures, leading to a problem that the feature extraction process and even the whole learning algorithm should be specifically redesigned for every different protocol. Besides, the manual feature extraction adds difficulty to achieve optimal performance.

An alternative method is to classify packets based on raw packet bytes rather than header fields. ML methods, especially neural networks, are also widely applied for this, such as [181–183]. These approaches have high accuracy and are not limited to specific protocols. However, they can only be deployed in a remote server/host rather than a data plane switch since it is difficult to define and install flow rules for each possible byte substring of the packet. Accessing the remote server adds extra latency and therefore packets cannot be processed at the line speed.

Learning Which Flow Rules to Install in SDN Switches
A third method that seeks to combine the merits of the two methods above was proposed in [184] in an IoT setting. The idea is to first train a neural network as a packet classifier taking as input the raw packet bytes, and in a later stage use the output of the neural network to select packet byte substrings as header fields and install appropriate flow rules in the data plane switch (IoT gateway). The two-stage process is illustrated in Figure 3.15a and the structure of the neural network in Figure 3.15b.

The rationale is that a packet classifier trained with the raw packet bytes (rather than specific headers) can achieve high accuracy without requiring to manually pre-define a set of headers as input features of the ML model and hence being able to handle different protocols with different packet header structures. Importantly, due to stage 2, there is no need to access the ML model in the remote server to do the inference and hence the packet classification, which would incur extra latency. Instead, packet classification can take place directly in the switches by installing appropriate flow rules for blocking the packets. The technical novelty of this approach lies in the way these rules are defined. By looking at the structure of the convolutional neural network, we can observe that different neurons correspond to different substrings of the packet bytes. The higher the layer the neuron belongs to the larger the corresponding substring. Hence,

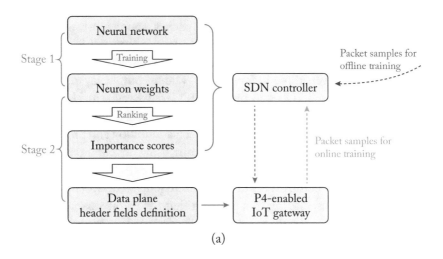

(a)

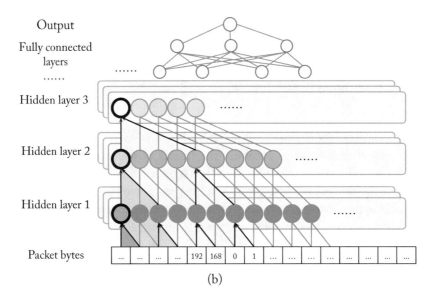

(b)

Figure 3.15: (a) Illustration of the two-stage learning approach in [184]. Packet classification is realized by the SDN control plane in Stage 1 by training a Neural Network (NN). Stage 2 takes place in the data plane switch that where header fields are defined based on the NN weights (importance scores) and flow rules matching these headers are installed. (b) Structure of the dilated convolutional neural network (Dilated CNN) trained in Stage 1.

the weight of each neuron in the trained model can be regarded as its importance score—how important the corresponding byte substring is for packet classification. Then, a small number of substrings with large weights are greedily picked as the packet headers and flow rules are defined based on them to mimic the packet classification logic of the neural network.

3.6   FAULT MANAGEMENT

Fault tolerance, reliability, and resilience to failures are important challenges in network operations. Usually, networks have to satisfy service level agreements (SLAs) specifying strict requirements on response time, packet losses, etc. Resiliency mechanisms are needed so that failures in the networks do not cause service disruptions and risk violating SLAs. While we covered the topic of failures in the control plane of the SDN network (the controller and the control channel) in the previous chapter, we did not discuss how to handle failures of network components in the data plane (links and switches). This section will cover this topic. We start with basic failure recovery methods of restoration and protection. We next describe solutions based on the concept of stateful data plane. We conclude this section with a discussion on optimization of these methods.

3.6.1   RESTORATION AND PROTECTION

For clarity, we use the term failure when the network element cannot deliver a service, whereas a fault is the root cause of the failure. The fault management process includes fault detection, fault localization and fault recovery. There are several mechanisms in traditional networks for handling link and node failures. At one level, when link failures are detected routing protocols automatically recompute paths and route along alternate paths. This requires the distributed routing protocols to distribute new link state information and for the protocol to converge to ensure consistency. For faster recovery, there are various local fast re-route mechanisms that are deployed and which can initiate immediate re-routing around failures without waiting for routing protocols to recompute paths. Note that the cause of failure may not be solely due to hardware failures and other factors such as configuration errors can be a problem as well.

SDN brings new opportunities and challenges to fault management. The programmability and flexibility in network management provided by SDN's open interfaces can be useful in automating fault management tasks. The centralized controller maintains a global view of the whole network and this can be useful in quickly isolating the affected switches and rerouting flows. The controller can react to these failures by removing affected flow entries and installing new entries in the affected switches so as to re-route flows away from the failed components in a globally optimized manner. Such re-routing is one approach to service *restoration*.

Figure 3.16a illustrates an example of restoration for a network of five OpenFlow switches A, B, C, D, and E. The SDN controller first installs a path A->B->C by installing flow rules at the three switches. When the controller receives a notification message that link BC has failed, it calculates the new path A->D->E->C. Then, it replaces the flow rule in switch A with a new

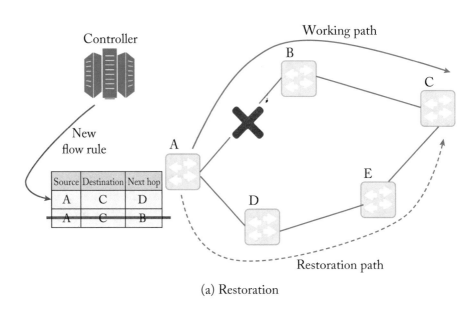

(a) Restoration

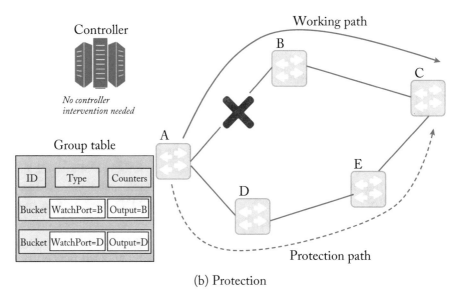

(b) Protection

Figure 3.16: Fault management in SDN.

one forwarding packets to next hop node D. The controller also needs to install new flow rules to the switches D, E, and C to realize flow routing over this path. From now on, the packets will be routed over the new path.

Despite the benefits, with this approach, restoration requires notifying the link failure event to the centralized controller with non-negligible delays and control overheads. For example, controller reaction times on the order of 100 ms have been measured. The restoration time also depends on the number of flows to be restored, path lengths, and may take even longer for larger networks. Importantly, there is the possibility in some types of networks that the controller and the switches may even be disconnected due to failures of the control channels, rendering impossible restoration with this approach until the connection is re-established. The issue of controller disconnection is of particular relevance in emerging applications of SDN in specialized networks of varying connectivity that we will discuss in the next chapter.

To avoid such problems, *protection* mechanisms that do not rely on the controller to re-route have been used as an alternative to controller based restoration. Here, failure recovery takes place completely in the data plane. The main idea is to pre-compute the alternative protection path and establish it together with the primary working path, as illustrated in Figure 3.16b. To achieve this, for instance, OpenFlow 1.1.0 used the so-called group table concept. Unlike the flow table, a group table contains group entries that each specifies a number of actions. Each Group Entry consists of a group ID, a group type, and a number of action buckets. An action bucket consists of a watchport and a set of actions that are to be executed based on the value of the associated watchport. For example, in the figure, the first action bucket describes what to do with the packet under normal conditions (i.e., forward from switch A to B). If this action bucket has been declared as unavailable that is due to change in status of a bucket (i.e., the link AB has failed), the packet is treated according to the next available bucket (i.e., forwarded through link AD).

3.6.2   STATEFUL DATA PLANE

A new wave of research considered the possibility of offloading some stateful traffic processing and control tasks from the controller to being directly inside the data plane. This approach, usually referred to as *stateful data plane*, was exploited in [185] to perform fast failover in the data plane in a similar yet more efficient way than OpenFlow 1.1.0 does. There are in fact several software implementations and platforms of stateful data plane and a comprehensive survey is available at [186]. Among them, OpenState has attracted increasing attention and has already been implemented as an OpenFlow 1.3 experimenter extension. The main idea in OpenState is that packets are matched in the flow table not only based on their header values but also based on a set of state values. The switches can keep track of these states and use them to make local decisions like re-routing packets to avoid failed links.

Figure 3.17 illustrates an example in which OpenState is used for fast re-routing of packets in the data plane. When a failure of a link in the primary path happens, the switch that cannot

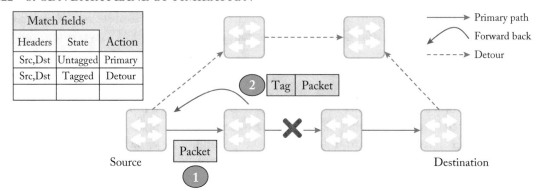

Figure 3.17: Example of failure recovery with OpenState. Upon failure, packets are tagged and forwarded back through the primary path to the source to trigger a state transition (from untagged to tagged). Source will forward all next packets directly through the detour path.

forward the packet to the next hop switch any more, first tags the packet (e.g., with a MPLS label containing information on the failure event) and then forwards it back through the primary path toward the source. The switch that receives the tagged packet processes it and performs a state transition so that all subsequent packets coming from the source switch will be matched on this new state and forwarded on the new path (detour). Note that the OpenFlow's local failover mechanism is not able to achieve such re-routing and would have to route all packets (not just the first packet) until the switch adjacent to the failed link and then re-route each of them back to the source. OpenState with its advanced tagging and state transition features can avoid unnecessary packet transmission overhead.

3.6.3   BACKUP PATH OPTIMIZATION

Regardless of whether group tables (as in OpenFlow 1.1.0) or states (as in OpenState) are adopted to realize the protection mechanism, a common challenge we face is how to select the path that will be used for re-routing each flow when the working path fails. We refer to the former path as backup path from now, also mentioned earlier as protection path or detour.

The centralized controller has to compute the backup paths and install the corresponding flow rules in advance, before the actual network failure happens. The more backup paths per flow that are computed the higher the level of protection provided but at the cost of more backup rules installed. In the simple case that at most one link failure can happen in the network at a time, a single backup path that is disjoint (without common links) of the primary path is enough to protect each flow. We can compute these paths by extending the multi-commodity flow problem, we described in previous sections, this time to decide both the working and backup paths. Specifically, we introduce an additional variable y_{fp}^b to denote that path p is used as backup for flow f. Then, the problem can be formulated as follows.

Backup Path Problem: $\min \theta$ (3.41)

$$s.t. \sum_{p \in P_f} y_{fp} = 1, \ \forall f \in \mathcal{F} \tag{3.42}$$

$$\sum_{p \in P_f} y_{fp}^b = 1, \ \forall f \in \mathcal{F} \tag{3.43}$$

$$\sum_{f, p \in P_f | e \in p} y_{fp} \lambda_f \leq \theta c_e, \ \forall e \in \mathcal{E} \tag{3.44}$$

$$y_{fp} + y_{fp'}^b \leq 1, \ \forall p, p' \in \mathcal{P} : p \cap p' \neq \emptyset \tag{3.45}$$

$$y_{fp}, \ y_{fp}^b \in \{0, 1\}, \ \forall f \in \mathcal{F}, p \in P_f. \tag{3.46}$$

In the above formulation, constraint (3.43) denotes that exactly one backup path will be picked for each flow, while constraint (3.45) ensures that the backup and working paths of a flow will be disjoint. Note here that the variables in constraint (3.46) are binary values and therefore routing is unsplittable. As in previously presented formulations, the objective is to minimize network congestion modeled as the maximum link utilization (variable θ).

In the above formulation, the congestion objective depends only on the traffic on the working paths, but not the backup paths. Indeed, constraint (3.44) aggregates only the traffic on the working paths. However, even a single link failure can trigger multiple flows to be re-routed to their backup paths. These paths (although disjoint to the primary path) may overlap one another and therefore congestion can happen in one or more links. Handling such cases is challenging especially for the case that multiple concurrent failures can happen. Besides, the total length of the paths (primary and backup) should be bounded to prevent excessive use of memory for flow rule installation.

3.7   NETWORK MEASUREMENT IN THE DATA PLANE

Network measurement is very important for various applications such as traffic engineering, scheduling, network slicing and security applications discussed above. While we covered this topic at the end of Chapter 2 (Section 2.2.2), we only considered the case that network measurement is performed by the control plane i.e., the SDN controller. The emergence of the programmable data plane concept, in the early 2010s [19, 20]. with its ability to potentially customize in-band telemetry in a programmable way catalyzed the offloading of many of the network measurement tasks from the controller to the data plane, thereby realizing more direct and effective network measurement. Furthermore, programmable in-band *network telemetry* became an area of active research interest. Moreover, network telemetry itself became terminology associated with the recent advanced monitoring methods.

3.7.1   IN-BAND NETWORK TELEMETRY

In-band network telemetry is representative of network telemetry and as mentioned earlier has been receiving increasing research attention [187]. It relies on probe packets with a variable-length label stack reserved in the packet header. The probe packets are periodically generated and injected into the network by in-band network telemetry agents, and they are queued and forwarded in the network like any other packet. As a probe packet is forwarded from one switch to another, it extracts switch-internal states such as link utilization or queue status. These meta-data are inserted into the label stack inside the packet. The process continues until the packet reaches its destination in which case the metadata will be sent to a remote centralized controller for further analysis. Although the controller still participates in the network measurement and will be the one who is ultimately responsible for the analysis, the data plane actively contributes to the process making the overall process more scalable. The metadata can be carried by independent probe packets separate from the user data packets, or it can be inserted inside and as part of the existing user data packets, as illustrated in Figure 3.18.

There are several challenges in realizing an in-band network telemetry system. First, for each probe packet injected into the network, we need to decide which network path it will cover. To achieve a network-wide view of the network, all the links should be covered and therefore multiple probe packets and corresponding paths are needed. Second, these probe packets will be generated periodically to keep the controller updated about the network state. Thus, the shorter that period is the more bandwidth will be consumed for carrying the probe packets. Third, the number of agents increases with the number of separate paths. Therefore, the controller needs to handle a higher telemetry workload sent by these agents. Fourth, at each period of this process, the controller will have to wait until it collects the probe packets from all paths before it forms the global network view. Consequently, if a path is much longer than the others, then it will delay the decision of the controller. Clearly, more balanced paths reduce this skew.

The work in [188] proposed two heuristic algorithms to determine a set of non-overlapping paths that cover the entire network. The first heuristic is based on the well-known depth-first search (DFS) algorithm. DFS is a recursive algorithm that uses the idea of backtracking. It involves exhaustive searches of all the vertices in a graph by going ahead, if possible, else by backtracking. The word backtrack refers to the case that you are moving forward and there are no more vertices along the current path you have not visited before, and so you have to move backward on the same path to find vertices to traverse. When applied to the in-band network telemetry problem, the idea is to create a separate path for a probe packet each time the algorithm is forced to backtrack. This path is built from the current sequence of vertices visited before backtracking. The process will end when all the edges in the graph are visited. By that time we will have extracted multiple non-overlapping paths covering the entire network. The second heuristic is more sophisticated and is specifically designed to optimize the number and the length of the paths (which are the two main limitations of the DFS algorithm). It is based on the definition of Euler trail which is a walk in a graph that visits every edge exactly once

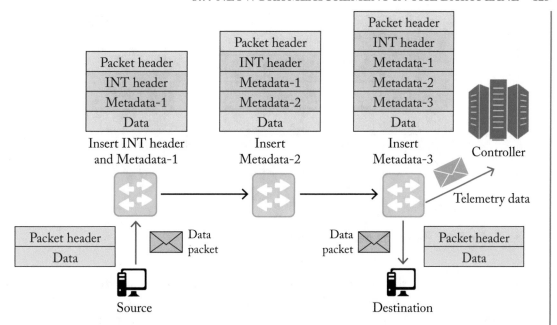

Figure 3.18: In-band network telemetry process. A user (source) transmits data packets to another user (destination) in the network. Telemetry metadata are inserted into the packet at each intermediate switch. Before the packet reaches its destination, the metadata are extracted and forwarded to the remote controller for further processing and analysis.

and which starts and ends on the same vertex. The algorithm exploits several theorems in graph theory about this trail. At a high level, the algorithm works by iteratively extracting from the graph a path between a pair of vertices with odd degrees until all the vertices are extracted from the original graph.

We remark that by picking non-overlapping paths, both the above algorithms reduce the bandwidth overheads of the process since they avoid redundant transmissions of the same telemetry metadata. Still, the amount of metadata and corresponding bandwidth overheads can be significant which challenges the scalability of the system for large networks. A promising way to deal with this problem could be to compress, filter or aggregate the metadata upon collection, an approach that has not been explored yet in the literature.

3.7.2   SKETCHES

An alternative network measurement method is using sketches. These are special data structures that use hash functions to summarize traffic statistics of flows, with bounded error and using small memory space. The sketches can be installed at the switches in the data plane to collect and measure the flows traversing the switches. Therefore, they represent another way to take

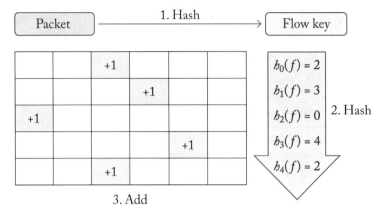

Figure 3.19: The count-min sketch example.

measurements in the data plane, different than by the metadata inserted in the packets by the in-band network telemetry methods. Sketches have been used for several measurement tasks such as detecting heavy hitters or estimating the flow size distribution in a network.

Figure 3.19 illustrates an example of a count-min sketch [189]. This type of sketch consists of a two-dimensional array with w columns and d rows. Every row is associated with a different hash function $h_i(f)$ that maps any flow f to a value in $\{0, 1, \ldots, w-1\}$. When a packet arrives, the sketch computes the hash value of its flow id for each hash function. The returned hash value is used as the index to pick an element of the array per row. The value of each of these elements is increased by 1. This way, we can estimate the size of each flow. Depending on the array size (w and d values), the estimation incurs an error which can be bounded with high probability.

As network management and applications become more complex and diverse, a single sketch is difficult for implementing all measurement tasks. Therefore, multiple sketches need to be deployed in the network. Deploying sketches in a network is not without challenges. To summarize the traffic, the sketches consume non-negligible computation power of the switch they run on. This can easily become the bottleneck as the network carries more and more traffic. To overcome this issue, an idea proposed in [190] is to balance the computing overhead of traffic measurement across all switches in the network. Since the computing overhead at a switch depends on the amount of traffic that passes through it and is measured by the sketch, the authors suggested to optimize the flow routing in the network together with sketch configuration. The latter means to decide which sketch(es) and on which switch will each flow be processed.

The joint traffic routing and sketch configuration problem is another extension of the multi-commodity flow problem presented in Section 3.2.2. The difference is that in addition to the routing variables y_{fp} for each flow f and path p, we also need to define the sketch configuration variables m_{fu}^s. This binary variable indicates whether flow f will be measured by sketch s at switch u or not. Then, the following key constraint needs to be added to the problem formulation:

$$\sum_{f \in \mathcal{F}} \sum_{s \in \mathcal{S}} m_{fu}^s N(f) P(s) \leq \theta p_u, \quad \forall u \in \mathcal{V}, \tag{3.47}$$

where $\mathcal{S}$ indicates the set of all available sketches. $N(f)$ denotes the number of packets of flow f. $P(s)$ stands for the average computing overhead per packet for sketch s, while p_u is the overall computation capacity of switch u. Since the objective of the problem is to minimize the variable θ (see Equation (3.8)), θ value represents the maximum computation load among all the switches.

The authors in [190] considered the unsplittable routing case of the problem where both the y_{fp} and m_{fu}^s are binary values. To solve the problem, they proposed a randomized rounding type of algorithm that, after solving a linear relaxation of the problem, rounds to binary values the m_{fu}^s variables randomly based on the fractional solution values of the relaxed problem. They further presented a primal-dual type of algorithm to deal with the online version of the problem where flows arrive to the network one-by-one. We remark that sketches can be regarded as a type of services, while the sketch configuration and routing problem can be regarded as a special case of the service chaining problem we discussed in Section 3.4. Therefore, the same service chaining algorithms already discussed can be used to solve this special case.

CHAPTER 4

Future Research Directions

Despite all the current work being done in SDN many significant research challenges remain to be addressed. In this section, we discuss these challenges and present future research directions. Following the same categorization as in the rest of the monograph, we discuss separately the challenges for the control and data planes of the SDN network. We continue with open questions pertaining to the application of ML to SDN, and conclude with emerging scenarios applying SDN to mobile and ad hoc network environments.

4.1 CHALLENGES IN THE CONTROL PLANE

Controller Placement

The existing literature on the controller placement problem focuses on a fully deployed SDN network. In practice, though, SDN will be deployed gradually over time and in incremental steps. The controller placement problem should be revisited by taking into account the architecture of hybrid SDN. Incremental deployment may require adaptive controller placement solutions. As the number of SDN switches increases in such hybrid networks, more controllers will be needed because of the increased control traffic and the likely increased geographic span. Adaptive controller placement solutions have been proposed to deal with topology changes in volatile networks of vehicles and drones. Solutions like these should be very useful for the hybrid SDN case too. Another open area for investigation in this context is the impact of the architecture of controllers. While controller placement has been extensively studied for the flat controller architecture, the alternative hierarchical architecture raises some new issues. Namely, there is a need to investigate the latency between the layers in the hierarchy of controllers and how this impacts the different applications. Higher-layer controllers may be more powerful and responsible for more complex applications with broader scope and interface with the lower-layer applications. Therefore, minimizing the impact of inter-controller latency can be significant for these applications. Besides, the resilience of these architectures where one controller can take over control of the applications when another controller in a different layer fails is an issue for further research.

Control Traffic Management

In a hybrid SDN deployment, the impact of traditional network elements on the control channel should also be carefully considered. These network elements may be of different types (e.g., router, firewall, encoder, etc.) and can affect the latency of the control channel. On the other hand, they can be also useful for managing the control traffic (e.g., filtering, prioritizing, etc.).

The controller synchronization problem should be revisited to account for these particular challenges and opportunities. Besides, the existing synchronization methods assume an eventually consistent protocol. It would also be interesting to model the problem for more complex protocols where controllers may synchronize in an event-driven rather than periodic manner. Furthermore, if the control channels are wireless they can interfere with each other as well as with the data channels (in case of in-band control). How to provision in-band and/or out-band control channels given the limited resources in the wireless network is another fascinating topic of research. Such problems become increasingly relevant as SDN applications in wireless networks become more and more popular, as we will also discuss in the sequel.

4.2   CHALLENGES IN THE DATA PLANE

SDN Deployment

While there has been much research on optimizing SDN deployment, most are mainly based on simplified network traffic models ignoring practical scenarios such as how the traffic changes between peak and off-peak hours during the day. Also, they assume rather simple deployment cost models (e.g., a total monetary budget for deployment). Other pragmatic constraints could also be introduced, such as the fact that devices at the edge of the network may be more amenable for SDN deployment than at the core due to operational and cost reasons. Another relevant question is how to collect topology information from traditional devices, which are not directly connected to the SDN controller. Emerging lightweight network monitoring technologies such as network telemetry can be used but have to be studied further in a hybrid SDN context.

Traffic Engineering

TE mechanisms have been a topic of research since the beginning of the Internet. The emergence of SDN fueled this research activity further and shifted the attention toward more sophisticated centralized algorithms that have now become easier to implement in practice. Most of the TE schemes proposed in an SDN context target the minimization of network congestion or related metrics. Of equal importance is using TE with resilience and fault management objectives as well. Additional failure recovery mechanisms should be implemented so that recovery can be achieved with low communication overhead and lesser or even no controller involvement, with the latter requiring the push of some more fault-management intelligence from the controller to the data plane switches. Besides, the compression/wildcard flow rule placement problem introduces intractable formulations that are far from being addressed. Although optimizing the use of routing tables has been studied in traditional networks before the advent of SDN, the capabilities of SDN make it more complex. Specifically, the existence of multiple packet header fields that can be used for matching increases the solution space of the problem. Also, the use of two different memory technologies for rule-matching (fast TCAM and slower random-access memory) motivates the development of optimized dual-memory rule placement methods.

4.3   MACHINE LEARNING APPLICATIONS OF SDN

Given the increasing popularity of data-driven ML methods in networking, we discuss here separately the challenges pertaining to its application to SDN. These involve both the control and the data plane sides of the network. ML methods have been applied to various networking problems such as intrusion detection (as we discussed in Section 3.5), network bottleneck detection [191], routing optimization [192], resource allocation [193], congestion control [194], and bitrate selection for video streaming [195]. The networking community had developed a large set of network models, algorithms, and optimization strategies to address these problems long before the ML methods appeared. The limitation of these traditional strategies is that we can optimize only what we can reasonably model. However, in many cases, especially in large and complex networks, no accurate models may exist or existing models may require making simplifying assumptions like assuming ideal properties or mathematically useful assumptions for real-world networks (e.g., generation of traffic with Poisson distribution). This motivates the application of ML to develop a new class of models—ML-based models that are particularly focused on difficult to model network behavior.

SDN makes it easier to apply ML techniques [196]. The SDN controller can use its network-wide view to collect network traffic statistics, analyze them, and train a corresponding ML model. Using this model, the controller can automatically learn how to make optimal decisions (e.g., for traffic routing) to adapt to network environments and enforce these decisions in the data plane by dynamically installing flow rules. Still, many challenges exist when designing such ML methods. First, it is unclear what is the right learning algorithm (supervised, unsupervised, reinforcement learning) to use and this may be different for different networking problems. Second, how exactly to design the learning algorithm is challenging. For instance, what should be the right input and output representation when building a neural network for automatically generating routing decisions? Third, where the learning model should reside can also be a design choice. The SDN controller appears to be the de facto choice but this is typically hosted in a server and therefore access to it for inferring optimal decisions and translating them into flow rules in the data plane can incur higher latencies than local placement at the switches. Below, we discuss in more detail two of these issues.

4.3.1   LEARNING TO ROUTE

While routing is a fundamental and well-investigated networking task, it still presents optimization challenges due to the dynamic and often unpredictable nature of traffic conditions. Typically, a network uses observations of the previous traffic conditions to optimize routing, with the need to adapt routing when traffic conditions deviate from the predicted traffic matrix. ML enables an alternative method of making routing optimization decisions using observed past traffic conditions to learn good routing configurations for future conditions.

There are two ways to define a learning model for routing. The first is rather intuitive and it is based on the fact that if the network conditions are known then we find the routing configurations by solving a multi-commodity flow problem (as we discussed in Section 3.2.2). Therefore, the learning model should be responsible for predicting the future traffic dynamics based on the past observations. The second way is less intuitive. It suggests training the learning model to find a mapping directly from the past traffic conditions to good routing configurations without explicit prediction of traffic. Both approaches exploit the fact that traffic often exhibits some level of regularity and therefore past traffic observations convey some knowledge about the future that can be used for routing decision optimizations.

Predicting future traffic conditions from past observations can be naturally represented as a supervised learning task. Preliminary results in this area, however, seem to indicate that supervised learning might be ineffective if the traffic conditions do not exhibit very high regularity [192]. Instead, learning directly the mapping from past traffic conditions to routing configurations appears more promising.

Another challenge is the large space of possible routing configurations that can be the output of the learning model. Consider, for example, the representation of routing decisions described in Section 3.2.2. This representation involves a number of variables equal to the number of network paths which in turn is exponential in the network size. Consequently, training a neural network would require a huge number of samples. A reinforcement learning algorithm would be too slow to converge. One way to expedite learning would be to pick a different more compact representation of routing. For instance, perform routing based on link weights so that the output of learning is just the weight vector of size equal to the number of links. Given the computed weights, a flow between a pair of source and destination nodes could be simply computed as the path of minimum aggregate weight [192]. The open question, however, is whether such representation limits the expressiveness of routing too much and how to balance this with the size of the output.

4.3.2   LEARNING IN THE DATA PLANE

Once the ML model is trained, the SDN controller can proactively install flow rules in switches to realize the corresponding network decisions and therefore handle packets directly in the data plane. While this approach alleviates the latency of reactively accessing the remote controller for inference of the ML model for future coming flows, the flow table size of some data plane switches is limited and may impose limitations.

An alternate approach with latency benefits is potentially offloading the inference task from the controller to the data plane elements. However, traditionally switches may not have the processing power to run a learning model by themselves and they rather need to rely on the controller for this as well as any other complex task.

The practice of centralizing inference may need to be revisited to better match the needs of ML applications in SDN. Such evolution of SDN can be facilitated by advances in both SDN and ML technologies. On the one hand, new high-level programming abstractions, as with P4, can enhance data plane flexibility. Further, lightweight neural network (NN) architectures like

Binarized Neural Networks (BNN) significantly reduce the computation and memory requirement for performing inference possibly to a level feasible for data plane switches in the future. A first attempt to apply these technologies for ML-assisted packet classification directly inside the switches was made in [198]. Extending these ideas for other network operations like routing and resource management is a challenging topic for future work.

4.4   MOBILE EXTENSIONS OF SDN

Last, we discuss emerging scenarios applying SDN to mobile ad hoc network environments. While most of the early work in SDNs has been related to wired networks, for instance data centers, a subsequent development is its use in wireless and mobile networks. For example, [199] proposed using SDN and shifting control functions from core gateways to middle boxes. This can eliminate management bottlenecks by decentralizing the network operation. On the other hand, [200] suggested the deployment of software defined radio access networks (RANs). The idea is to assign the management of multiple base stations to a global controller and, with this unified control, improve performance. An interesting point is the suggestion of splitting the control decisions into those requiring full information (hence assigned to the global controller) and those that need fast response (assigned to the radio elements). Going a step further in the data plane, [201] proposed to turn the base stations into fully programmable nodes, thus facilitating the virtualization of network resources.

The above developments were early concrete steps that have been made toward designing softwarized wireless and mobile networks. Interestingly, there have also been efforts to use SDN software switches even in handheld devices to programamtically control their uplink or device-to-device link transmissions via various interfaces (WiFi, Bluetooth) [202], [203]. This shows the potential for deploying SDN even in areas without fixed wireless infrastructure and toward SDN-enabled mobile ad hoc network (MANETs) architectures. However, this area of research is still being explored. While these ad-hoc networks are not of primary interest in the industry (where the prime interest is SDN in the context of 5G networks for flexible network operations, customization, and management), the research interest is driven by military and other special-purpose networks seeking to explore the potential of SDN to re-design their tactical MANETs formed in a battlefield environment, enabling new services and capabilities for the mobile tactical units.[1]

Consider, for example, the SDN-enabled MANET for a tactical network as in Figure 4.1. Here, two local SDN controllers are installed on portable stations that are in proximity to the mobile nodes. Each of them is able to view only a part of the network graph, and has to collect and send the respective network state information to the higher-level SDN controller located at a cloudlet. This global controller will construct the universal graph topology, devise the operation policies, and disseminate them to the forwarding nodes through the local controllers. The latter

---

[1]See, for example, the DAIS ITA research program funded by ARL [204].

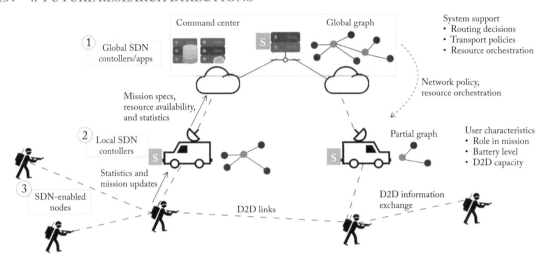

Figure 4.1: Overview of SDN-enabled MANET architecture for a tactical network. SDN local controllers (denoted by S) receive the network policies from the cloudlet (global controller) and configure the end nodes under their supervision accordingly.

can serve as bridges, or even take full control when the infrastructure connection is disrupted or the data path queries need to be serviced in real time.

There are many open issues that we need to address before we have a realizable implementation of an SDN-enabled MANET architecture as the one described above. A first challenge is to determine the SDN controller architecture. This requires deciding the placement of controllers in the network. Toward this goal, we could apply the same methods described in Section 2.1 to optimize metrics such as latency and quality of service. However, the network topology in this scenario is highly volatile with mobile nodes being constantly at risk of losing connectivity from their associated controllers. Therefore, neither the static methods nor the adaptive to traffic variation methods we described in Section 2.1.3 are applicable. Instead, an adaptive controller placement strategy that takes into consideration the mobility patterns of the nodes and the stability of the network links should be preferred.

While the aforementioned mobility-aware adaptive controller placement strategies can help when mobile nodes are disconnected from their controllers, they cannot by themselves address the problem completely. As we explained in Section 2.1, activating/deactivating controllers and migrating switches among them takes time and risks service interruption. Besides, it may not be always possible to activate a new controller to handle all possible mobility and network fragmentation scenarios and in the worse case this would require placing a controller at every possible node.

An alternative promising direction for handling network dynamism and controller disconnections is to delegate some level of control to data plane nodes. This can be realized by pre-computing and pushing to data plane nodes locally executable code fragments that realize control functions in an autonomous way, even when the controller is not reachable. This would of course revisit SDN's main philosophy of separating control and data planes. Yet, it is in line with high-function hardware and software switches that offer much more sophisticated support for persistent state and control functions in the data plane, without always querying a central controller. For example, if the data plane nodes are enhanced with the ability to perform stateful forwarding actions, as in P4 or OpenState, these control functions can take the form of state-dependent forwarding rules. Nevertheless, managing stateful, distributed systems efficiently and correctly is recognized as a long-standing challenging problem.

Another design choice is using hybrid architectures where control functions reside at both the central controller and the distributed data plane nodes. A rather straightforward design is to enable data plane nodes to run a legacy MANET protocol (e.g., OLSR or AODV) which, as an example of a simple local control function, can be used to locally re-route traffic through alternative network paths when link failures are detected. This control function can be programmed into stateful SDN switches where the state maintained by the switches is simply the link state information disseminated among nodes by the MANET protocol. In this scenario, the system should be able to detect disconnection from the controller and dynamically activate the MANET protocol as well as later roll back to the central controller mode of operation. However, such a design is rather opportunistic and best-effort, and carries with it all the limitations of MANET protocols (long convergence time, constrained to shortest paths, lacking capabilities for more advanced tasks such as link load balancing). Designing more advanced hybrid centralized/distributed control functions for mobile ad-hoc networks still has many open research challenges.

While the concept of SDN-enabled MANETs gives rise to several new problems that are of interest from a theoretical (modeling and optimization) point of view, such as those listed above, it also raises some important system questions. Specifically, it is not clear whether it is even feasible or how exactly SDN components (controller modules like OpenDaylight and ONOS and data path modules like Open vSwitch) can be hosted and run in existing mobile devices. The work in [59] made a first attempt to answer this question. This work deployed a testbed of an SDN-enabled wireless system using off-the-shelf mobile devices, as illustrated in Figure 4.2a. Specifically, the testbed consists of Nexus 4 Android smartphones that form a wireless network. The first smartphone works as an access point (hotspot) to provide the remaining three smartphones with WiFi connections. This represents a common wireless network scenario, where a node can either establish multihop connections to backbone networks, or exchange data with its peer in a device-to-device fashion. Besides, smartphones are representative of devices widely used in wireless networks. Due to their computation and storage constraints, this testbed illustrates a challenging scenario that is worthy of investigation.

Several steps are taken to make the network SDN-enabled. In each smartphone, the *chroot* environment is used to install the Ubuntu system running with its original Android system at

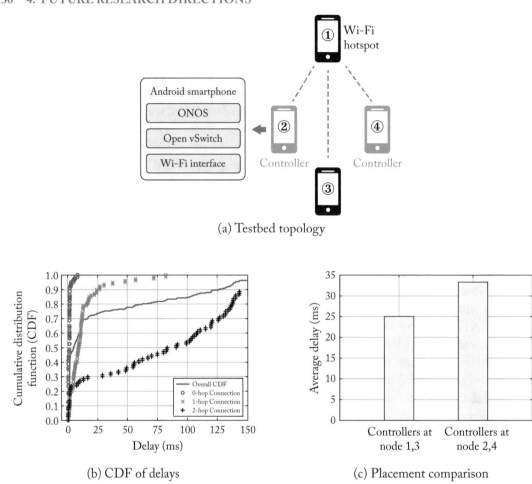

(a) Testbed topology

(b) CDF of delays

(c) Placement comparison

Figure 4.2: (a) Testbed of a multi-controller wireless system built from smartphones that enable Open vSwitch and ONOS. (b) CDF of controller-switch communication latencies. (c) Comparison of average latencies under different controller placements.

the same time. By this, we can install popular SDN-related software in the smartphone. First, Open vSwitch is installed to help make all devices work as data plane nodes (Open vSwitch adds virtual switching to the smartphone so enabling SDN support). Second, SDN controllers are deployed in Nodes 2 and 4. Though constrained in resources, the smartphone is still capable enough to run a controller instance, such as ONOS, which permits multiple controllers to work together in the form of a controller cluster (flat architecture). Next, a connection between the two controllers is established and each smartphone is assigned to its nearest controller with respect to the hop count length distance metric.

Measurements are taken on the network delay between the controller and data plane nodes. The devices are placed in a lab environment such that the distance over each wireless link is 2 meters. One frequent and important interaction between a controller and a data plane node is the request and response for flow statistics. Therefore, the latency at controller nodes is measured by analyzing the interval between sending such an OpenFlow request message and receiving its corresponding reply. Figure 4.2b shows the cumulative distribution function of the latencies for 250 measurements. The average value is tens of milliseconds which is higher than the latency reported in typical wired networks [24]. From the CDF plot, we notice that the variance is large, corresponding to the relatively unstable wireless links. It is common for the latency to go even beyond 100 milliseconds.

We also notice that the placement of controllers is important, because the latency depends highly on the distance between the data plane node and its controller. For example, Nodes 2 and 4 contain controllers locally, while Nodes 1 and 3 have one-hop and two-hop connections to the controller, respectively. As a result, as shown separately in Figure 4.2b, local connection shows almost zero latencies while the one-hop and two-hop connections show notable latencies. To further demonstrate this, we move ONOS controllers from Nodes 2 and 4 to Nodes 1 and 3. If we still assign each data plane node to its nearest controller, we observe that the average latency is 25% lower, as is depicted in Figure 4.2c.

The above experimental results demonstrate that modern edge network devices (smartphones) can act as controllers to manage other devices. The management latency highly depends on the number of wireless hops and can significantly change for different controller placement strategies (up to 25% difference in the above testbed).

While the feasibility of SDN-enabled mobile ad hoc networks has been shown, experiments with a larger number of devices are needed to show the scalability and limits of such deployments. More importantly, we need to evaluate mechanisms for dealing with disconnections of the controller from the data plane switches to ensure that network applications are not affected. Overall, the SDN architecture should evolve to become more robust and adaptive to the high level of dynamism in the mobile ad hoc networks.

Bibliography

[1] P. Cipresso, I. A. C. Giglioli, M. A. Raya, and G. Riva. The past, present, and future of virtual and augmented reality research: A network and cluster analysis of the literature. *Front Psychol.*, 9:2086, 2018. DOI: 10.3389/fpsyg.2018.02086. 1

[2] L. Da Xu, W. He, and S. Li. Internet of Things in industries: A survey. *IEEE Trans. Ind. Informat.*, 10(4):2233–2243, November 2014. DOI: 10.1109/tii.2014.2300753. 1

[3] A. Abhashkumar, J.-M. Kang, S. Banerjee, A. Akella, Y. Zhang, and W. Wu. Supporting diverse dynamic intent-based policies using Janus. *Proc. of the 13th International Conference on Emerging Networking Experiments and Technologies (CoNEXT'17). Association for Computing Machinery*, pages 296–309, New York, 2017. DOI: 10.1145/3143361.3143380. 1

[4] N. McKeown, T. Anderson, H. Balakrishnan, G. Parulkar, L. Peterson, J. Rexford, S. Shenker, and J. Turner. OpenFlow: Enabling innovation in campus networks. *SIGCOMM Comput. Commun. Rev.*, 38(2):69–74, 2008. DOI: 10.1145/1355734.1355746. 1, 3

[5] S. Jain, A. Kumar, S. Mandal, J. Ong, L. Poutievski, A. Singh, S. Venkata, J. Wanderer, J. Zhou, M. Zhu, J. Zolla, U. Hölzle, S. Stuart, and A. Vahdat. B4: Experience with a globally-deployed software defined wan. *Proc. of the ACM SIGCOMM Conference on SIGCOMM. Association for Computing Machinery*, pages 3–14, New York, 2013. DOI: 10.1145/2486001.2486019. 1

[6] Z. Zaidi, V. Friderikos, Z. Yousaf, S. Fletcher, M. Dohler, and H. Aghvami. Will SDN be part of 5G?. *IEEE Commun. Surv. Tutor.*, 52(4):3220–3258, 2018. DOI: 10.1109/comst.2018.2836315. 1

[7] B. A. A. Nunes, M. Mendonca, X. Nguyen, K. Obraczka, and T. Turletti. A survey of software-defined networking: Past, present, and future of programmable networks. *IEEE Commun. Surv. Tutor.*, 16(3):1617–1634, 2014. DOI: 10.1109/surv.2014.012214.00180. 2

[8] F. A. Lopes, M. Santos, R. Fidalgo, and S. Fernandes. A software engineering perspective on SDN programmability. *IEEE Commun. Surv. Tutor.*, 18(2):1255–1272, 2016. DOI: 10.1109/comst.2015.2501026. 2

[9] J. C. C. Chica, J. C. Imbachi, and J. F. B. Vega. Security in SDN: A comprehensive survey. *J. Netw. Comput. Appl.*, 159(102595), June 2020. DOI: 10.1016/j.jnca.2020.102595. 2

[10] N. Feamster, J. Rexford, and E. Zegura. The road to SDN: An intellectual history of programmable networks. *SIGCOMM Comput. Commun. Rev.*, 44(2):87–98, 2014. DOI: 10.1145/2602204.2602219. 2

[11] N. Anerousis, P. Chemouil, A. A. Lazar, N. Mihai, and S. B. Weinstein. The origin and evolution of open programmable networks and SDN. *IEEE Commun. Surv. Tutor.*. DOI: 10.1109/comst.2021.3060582. 2

[12] K. Psounis. Active networks: Applications, security, safety, and architectures. *IEEE Commun. Surv.*, 2(1):2–15, 1999. DOI: 10.1109/comst.1999.5340509. 3

[13] M. Caesar, N. Feamster, J. Rexford, A. Shaikh, and J. van der Merwe. Design and implementation of a routing control platform. *2nd USENIX NSDI*, 2005. 3

[14] T. Lakshman, T. Nandagopal, R. Ramjee, K. Sabnani, and T. Woo. The SoftRouter architecture. *3rd ACM Workshop Hot Topics Netw.*, 2004. 3

[15] M. Casado, M. J. Freedman, J. Pettit, J. Luo, N. McKeown, and S. Shenker. Ethane: Taking control of the enterprise. *Proc. of the Conference on Applications, Technologies, Architectures, and Protocols for Computer Communications (SIGCOMM'07). Association for Computing Machinery*, pages 1–12, New York, 2007. DOI: 10.1145/1282380.1282382. 3

[16] www.opennetworking.org 4

[17] www.ietf.org 4

[18] N. Gude, T. Koponen, J. Pettit, B. Pfaff, M. Casado, N. McKeown, and S. Shenker. NOX: Towards an operating system for networks. *SIGCOMM Comput. Commun. Rev.*, 38(3):105–110, 2008. DOI: 10.1145/1384609.1384625. 3

[19] P. Bosshart, D. Daly, G. Gibb, M. Izzard, N. McKeown, J. Rexford, C. Schlesinger, D. Talayco, A. Vahdat, G. Varghese, and D. Walker. P4: Programming protocol-independent packet processors. *SIGCOMM Comput. Commun. Rev.*, 44(3):87–95, July 2014. DOI: 10.1145/2656877.2656890. 4, 123

[20] H. Song. Protocol-oblivious forwarding: Unleash the power of SDN through a future-proof forwarding plane. *Proc. of the 2nd ACM SIGCOMM Workshop on Hot Topics in Software Defined Networking (HotSDN'13). Association for Computing Machinery*, pages 127–132, New York, 2013. DOI: 10.1145/2491185.2491190. 4, 123

[21] C. G. Miller. Telephone Switching System, U.S. Patent 3,189,687A, June 15, 1965. 4

[22] A. G. Fraser. Spider—an experimental data communications system. *Proc. IEEE Conference on Communications*, page 21F, June 1974. 4

[23] F. Bannour, S. Souihi, and A. Mellouk. Distributed SDN control: Survey, taxonomy, and challenges. *IEEE Commun. Surv. Tutor.*, 20(1):333–354, 2018. DOI: 10.1109/comst.2017.2782482. 7, 9

[24] B. Heller, R. Sherwood, and N. McKeown. The controller placement problem. *SIGCOMM Comput. Commun. Rev.*, 42(4):473–478, 2012. DOI: 10.1145/2377677.2377767. 10, 13, 14, 17, 18, 30, 33, 137

[25] T. Das, V. Sridharan, and M. Gurusamy. A survey on controller placement in SDN. *IEEE Commun. Surv. Tutor.*, 22(1):472–503, 2020. DOI: 10.1109/COMST.2019.2935453.

[26] R. Muñoz et al. Integrated SDN/NFV management and orchestration architecture for dynamic deployment of virtual SDN control instances for virtual tenant networks. *J. Opt. Commun. Netw.*, 7(11):62–70, 2015. DOI: 10.1364/JOCN.7.000B62.

[27] J. Zhao, H. Qu, J. Zhao, Z. Luan, and Y. Guo. Towards controller placement problem for software-defined network using affinity propagation. *Electron. Lett.*, 2017. DOI: 10.1049/el.2017.0093. 14, 15

[28] G. Wang, Y. Zhao, J. Huang, Q. Duan, and J. Li. A k-means based network partition algorithm for controller placement in software defined network. *Communications (ICC), IEEE International Conference on*, pages 1–6, 2016. DOI: 10.1109/icc.2016.7511441. 14, 15

[29] G. Wang, Y. Zhao, J. Huang, and Y. Wu. An effective approach to controller placement in software defined wide area networks. *IEEE Trans. Netw. Serv. Manage.*, 15(1):344–355, 2018. DOI: 10.1109/tnsm.2017.2785660. 14, 15

[30] M. Jarschel et al. Modeling and performance evaluation of an OpenFlow architecture. *Proc. 23rd Int. Teletraffic Congr.*, pages 1–7, 2011.

[31] B. P. R. Killi, E. A. Reddy, and S. V. Rao. Cooperative game theory based network partitioning for controller placement in SDN. *10th International Conference on Communication Systems and Networks (COMSNETS)*, pages 105–112, Bengaluru, 2018. DOI: 10.1109/comsnets.2018.8328186. 14, 16

[32] G. Yao, J. Bi, Y. Li, and L. Guo. On the capacitated controller placement problem in software defined networks. *IEEE Commun. Lett.*, 18(8):1339–1342, 2014. DOI: 10.1109/lcomm.2014.2332341. 14, 17

[33] C. Gao, H. Wang, F. Zhu, L. Zhai, and S. Yi. A particle swarm optimization algorithm for controller placement problem in software defined network. *International Conference on Algorithms and Architectures for Parallel Processing*, pages 44–54, Springer, 2015. DOI: 10.1007/978-3-319-27137-8_4. 14, 17

[34] S. Liu, H. Wang, S. Yi, and F. Zhu. NCPSO: A solution of the controller placement problem in software defined networks. *International Conference on Algorithms and Architectures for Parallel Processing*, pages 213–225, Springer, 2015. DOI: 10.1007/978-3-319-27137-8_17. 14, 17

[35] T. Zhang, P. Giaccone, A. Bianco, and S. D. Domenico. The role of the inter-controller consensus in the placement of distributed SDN controllers. *Comput. Commun.*, 113(15):1–3, 2017. DOI: 10.1016/j.comcom.2017.09.007. 14, 17

[36] A. Ksentini, M. Bagaa, T. Taleb, and I. Balasingham. On using bargaining game for optimal placement of SDN controllers. *Communications (ICC), IEEE International Conference on*, pages 1–6, 2016. DOI: 10.1109/icc.2016.7511136. 14, 17

[37] Y. Hu, W. Wendong, X. Gong, X. Que, and C. Shiduan. Reliability-aware controller placement for software-defined networks. *IFIP/IEEE International Symposium on Integrated Network Management (IM)*, 2013. 19, 20, 24, 29

[38] Y. Hu, W. Wang, X. Gong, X. Que, and S. Cheng. On reliability-optimized controller placement for software-defined networks. *China Commun.*, 11(2):38–54, February 2014. DOI: 10.1109/cc.2014.6821736. 19, 20, 24, 29

[39] M. Guo and P. Bhattacharya. Controller placement for improving resilience of software-defined networks. *4th International Conference on Networking and Distributed Computing*, 2013. DOI: 10.1109/icndc.2013.15. 20, 21, 30

[40] J. Liug, J. Liu, and R. Xie. Reliability-based controller placement algorithm in software defined networking. *Comput. Sci. Inf. Syst.*, 13(2):547–560, 2016. DOI: 10.2298/csis160225014l. 20, 22, 25

[41] F. Ros and P. Ruiz. Five nines of southbound reliability in software-defined networks. *HotSDN*, 2014. DOI: 10.1145/2620728.2620752. 20, 22, 23, 28, 29

[42] F. J. Ros and P. M. Ruiz. On reliable controller placements in software-defined networks. *Comput. Commun.*, 77:41–51, 2016. DOI: 10.1016/j.comcom.2015.09.008. 20, 22, 23, 28, 29

[43] L. Li, N. Du, H. Liu, R. Zhang, and C. Yan. Towards robust controller placement in software-defined networks against links failure. *IFIP/IEEE Symposium on Integrated Network and Service Management (IM)*. pages 216–223, Arlington, VA, 2019. 20, 23, 24

[44] Q. Zhong, Y. Wang, W. Li, and X. Qiu. A min-cover based controller placement approach to build reliable control network in SDN. *NOMS IEEE/IFIP Network Operations and Management Symposium*, pages 481–487, Istanbul, 2016. DOI: 10.1109/noms.2016.7502847. 20, 24

[45] L. F. Müller, R. R. Oliveira, M. C. Luizelli, L. P. Gaspary, and M. P. Barcellos. Survivor: An enhanced controller placement strategy for improving SDN survivability. *IEEE Global Communications Conference*, 2014. DOI: 10.1109/glocom.2014.7037087. 20, 25

[46] D. Hock et al. Pareto-optimal resilient controller placement in SDN-based core networks. *Teletraffic Congress (ITC), 25th International IEEE*, 2013. DOI: 10.1109/itc.2013.6662939. 20, 25, 26, 27, 29

[47] S. Lange, S. Gebert, T. Zinner, P. Tran-Gia, D. Hock, M. Jarschel, and M. Hoffmann. Heuristic approaches to the controller placement problem in large scale SDN networks. *IEEE Trans. Netw. Serv. Manage.*, 12(1):4–17, 2015. DOI: 10.1109/tnsm.2015.2402432. 20, 25, 26, 29

[48] B. P. R. Killi and S. V. Rao. Optimal model for failure foresight capacitated controller placement in software-defined networks. *IEEE Commun. Lett.*, 20(6):1108–1111, 2016. DOI: 10.1109/lcomm.2016.2550026. 20, 27, 28

[49] B. P. R. Killi and S. V. Rao. Capacitated next controller placement in software defined networks. *IEEE Trans. Netw. Serv. Manage.*, 14(3):514–527, 2017. DOI: 10.1109/tnsm.2017.2720699. 20, 28

[50] M. Tanha, D. Sajjadi, R. Ruby, and J. Pan. Capacity-aware and delay-guaranteed resilient controller placement for software-defined WANs. *IEEE Trans. Netw. Serv. Manage.*, 15(3):991–1005, September 2018. DOI: 10.1109/tnsm.2018.2829661. 20, 28

[51] P. Vizarreta, C. M. Machuca, and W. Kellerer. Controller placement strategies for a resilient SDN control plane. *8th International Workshop on Resilient Networks Design and Modeling (RNDM)*, pages 253–259, Halmstad, 2016. DOI: 10.1109/rndm.2016.7608295. 20, 29, 31

[52] Y. Jiménez, C. Cervelló-Pastor, and A. J. García. On the controller placement for designing a distributed SDN control layer. *IFIP Networking Conference*, 2014. DOI: 10.1109/ifipnetworking.2014.6857117. 20, 30

[53] T. Das and M. Gurusamy. Controller placement for resilient network state synchronization in multi-controller SDN. *IEEE Commun. Lett.*, 24(6), 2020. DOI: 10.1109/lcomm.2020.2979072. 20, 30

[54] A. Sallahi and M. St-Hilaire. Optimal model for the controller placement problem in software defined networks. *IEEE Commun. Lett.*, 19(1):30–33, 2015. DOI: 10.1109/lcomm.2014.2371014. 31, 32

[55] A. Sallahi and M. St-Hilaire. Expansion model for the controller placement problem in software defined networks. *IEEE Commun. Lett.*, 21(2):274–277, 2017. DOI: 10.1109/lcomm.2016.2621746. 31, 32

[56] Z. Zhao and B. Wu. Scalable SDN architecture with distributed placement of controllers for wan. *Concurre. Computat. Pract. Exper.*, 29(16), 2017. DOI: 10.1002/cpe.4030. 31, 32

[57] Y. Hu, T. Luo, N. C. Beaulieu, and C. Deng. The energy-aware controller placement problem in software defined networks. *IEEE Commun. Lett.*, 21(4):741–744, 2017. DOI: 10.1109/lcomm.2016.2645558. 31, 32

[58] Z. Su and M. Hamdi. MDCP: Measurement-aware distributed controller placement for software defined networks. *IEEE 21st International Conference on Parallel and Distributed Systems (ICPADS)*, 2015. DOI: 10.1109/icpads.2015.55. 31, 32

[59] Q. Qin, K. Poularakis, G. Iosifidis, and L. Tassiulas. SDN controller placement at the edge: Optimizing delay and overheads. *IEEE International Conference on Computer Communications (Infocom)*, 2018. DOI: 10.1109/infocom.2018.8485963. 31, 33, 58, 135

[60] Q. Qin, K. Poularakis, G. Iosifidis, S. Kompella, and L. Tassiulas. SDN controller placement with delay-overhead balancing in wireless edge networks. *IEEE Trans. Netw. Serv. Manage.*, 15(4):1446–1459, December 2018. DOI: 10.1109/tnsm.2018.2876064. 31, 33, 58

[61] M. F. Bari, A. R. Roy, S. R. Chowdhury, Q. Zhang, M. F. Zhani, R. Ahmed, and R. Boutaba. Dynamic controller provisioning in software defined networks. *Proc. of the 9th International Conference on Network Service Management (CNSM)*, 2013. DOI: 10.1109/cnsm.2013.6727805.

[62] M. T. I. ul Huque, W. Si, G. Jourjon, and V. Gramoli. Large-scale dynamic controller placement. *IEEE Trans. Netw. Serv. Manage.*, 14(1):63–76, 2017. DOI: 10.1109/tnsm.2017.2651107.

[63] H. K. Rath, V. Revoori, S. M. Nadaf, and A. Simha. Optimal controller placement in software defined networks (SDN) using a non-zero-sum game. *Proc. of IEEE International Symposium on a World of Wireless, Mobile and Multimedia Networks*, 2014. DOI: 10.1109/wowmom.2014.6918987.

[64] M. J. Abdel-Rahman, E. A. Mazied, A. MacKenzie, S. Midkiff, M. R. Rizk, and M. El-Nainay. On stochastic controller placement in software-defined wireless networks. *IEEE Wireless Communications and Networking Conference (WCNC)*, 2017. DOI: 10.1109/wcnc.2017.7925942.

[65] K. Sudheera, K. Liyanagea, M. Maa, and P. H. J. Chong. Controller placement optimization in hierarchical distributed software defined vehicular networks. *Comput. Netw.*, 135:226–239, 2018. DOI: 10.1016/j.comnet.2018.02.022.

[66] T. Benson, A. Akella, and D. Maltz. Network traffic characteristics of data centers in the wild. *IMC*, 2010. DOI: 10.1145/1879141.1879175. 33

[67] A. Dixit, F. Hao, S. Mukherjee, T. V. Lakshman, and R. R. Kompella. ElastiCon: An elastic distributed SDN controller. *Proc. ACM/IEEE Symp. Archit. Netw. Commun. Syst. (ANCS)*, pages 17–27, October 2014. DOI: 10.1145/2658260.2658261. 34, 42

[68] H. C. Cheng, Z. Wang and S. Chen. DHA: Distributed decisions on the switch migration toward a scalable SDN control plane. *IFIP Networking Conference (IFIP Networking)*, pages 1–9, IEEE, 2015. DOI: 10.1109/ifipnetworking.2015.7145319. 34, 35, 36, 40

[69] G. Cheng, H. Chen, H. Hu, and J. Lan. Dynamic switch migration towards a scalable SDN control plane. *Int. J. Commun. Syst.*, 29(9):1482–1499, 2016. DOI: 10.1002/dac.3101. 35, 36

[70] T. Wang, F. Liu, J. Guo, and H. Xu. Dynamic SDN controller assignment in data center networks: Stable matching with transfers. *IEEE INFOCOM—The 35th Annual IEEE International Conference on Computer Communications*, pages 1–9, San Francisco, CA, 2016. DOI: 10.1109/infocom.2016.7524357. 35, 37

[71] Y. Kyung, K. Hong, T. M. Nguyen, S. Park, and J. Park. A load distribution scheme over multiple controllers for scalable SDN. *7th International Conference on Ubiquitous and Future Networks*, pages 808–810, Sapporo, 2015. DOI: 10.1109/icufn.2015.7182654. 35, 37

[72] Y. Hu, W. Wang, X. Gong, X. Que, and S. Cheng. Balanceflow: Controller load balancing for openflow networks. *Proc. IEEE CCIS*, pages 780–785, Hangzhou, China, October 2012. DOI: 10.1109/ccis.2012.6664282. 35, 38

[73] H. Selvi, G. Gür, and F. Alagöz. Cooperative load balancing for hierarchical SDN controllers. *Proc. IEEE HPSR*, pages 100–105, Yokohama, Japan, June 2016. DOI: 10.1109/hpsr.2016.7525646. 35, 38

[74] V. Sridharan, M. Gurusamy, and T. Truong-Huu. On multiple controller mapping in software defined networks with resilience constraints. *IEEE Commun. Lett.*, 21(8):1763–1766, August 2017. DOI: 10.1109/lcomm.2017.2696006. 35, 38

[75] Y. Zhou, Y. Wang, J. Yu, J. Ba, and S. Zhang. Load balancing for multiple controllers in SDN based on switches group. *19th Asia-Pacific Network Operations and Management Symposium (APNOMS)*, pages 227–230, Seoul, 2017. DOI: 10.1109/apnoms.2017.8094139. 35, 38, 39

[76] C. Wang, B. Hu, S. Chen, D. Li, and B. Liu. A switch migration-based decision-making scheme for balancing load in SDN. *IEEE Access*, 5:4537–4544, 2017. DOI: 10.1109/access.2017.2684188. 35, 39

[77] T. Hu, J. Lan, J. Zhang et al. EASM: Efficiency-aware switch migration for balancing controller loads in software-defined networking. *Peer-to-Peer Netw. Appl.*, 12:452–464, 2019. DOI: 10.1007/s12083-018-0632-6. 35, 39

[78] Y. Xu, M. Cello, I.-C. Wang, A. Walid, G. Wilfong, C. H.-P. Wen, M. Marchese, and H. J. Chao. Dynamic switch migration in distributed software-defined networks to achieve controller load balance. *IEEE J. Sel. Areas Commun.*, 37(3):515–529, February 2019. DOI: 10.1109/jsac.2019.2894237. 35, 39, 40, 41

[79] M. F. Bari, A. R. Roy, S. R. Chowdhury, Q. Zhang, M. F. Zhani, R. Ahmed, and R. Boutaba. Dynamic controller provisioning in software defined networks. *Netw. Serv. Manage. (CNSM), 9th International Conference on IEEE*, pages 18–25, 2013. DOI: 10.1109/cnsm.2013.6727805. 43, 44

[80] M. He, A. Basta, A. Blenk, and W. Kellerer. Modeling flow setup time for controller placement in SDN: Evaluation for dynamic flows, 2017. DOI: 10.1109/icc.2017.7996654. 43, 44

[81] M. T. I. Ul Huque, G. Jourjon, and V. Gramoli. Revisiting the controller placement problem. *Local Computer Networks (LCN), IEEE 40th Conference on*, pages 450–453, 2015. DOI: 10.1109/LCN.2015.7366350. 43, 44

[82] H. K. Rath, V. Revoori, S. Nadaf, and A. Simha. Optimal controller placement in software defined networks (SDN) using a non-zero-sum game. *World of Wireless, Mobile and Multimedia Networks (WoWMoM), IEEE 15th International Symposium on*, pages 1–6, 2014. DOI: 10.1109/wowmom.2014.6918987. 43, 45

[83] X. Lyu, C. Ren, W. Ni, H. Tian, R. P. Liu, and Y. J. Guo. Multi-timescale decentralized online orchestration of software-defined networks. *IEEE J. Sel. Areas Commun.*, 36(12):2716–2730, December 2018. DOI: 10.1109/jsac.2018.2871310. 43, 45, 46, 48

[84] J. Yang, X. Yang, Z. Zhou, X. Wu, T. Benson, and C. Hu. FOCUS: Function offloading from a controller to utilize switch power. *Proc. of IEEE NFV-SDN'16*, 2016. DOI: 10.1109/nfv-sdn.2016.7919498. 48

[85] X. Huang, S. Bian, Z. Shao, and Y. Yang. Predictive switch-controller association and control devolution for SDN systems. *Proc. of IEEE/ACM IWQoS*, 2019. DOI: 10.1145/3326285.3329070. 48

[86] F. Akyildiz, S.-C. Lin, and P. Wang. Wireless software-defined networks (W-SDNs) and network function virtualization (NFV) for 5G cellular systems: An overview and qualitative evaluation. *Comput. Netw.*, 93(12):66–79, 2015. DOI: 10.1016/j.comnet.2015.10.013. 49

[87] M. J. Abdel-Rahman, E. A. Mazied, A. MacKenzie, S. Midkiff, M. R. Rizk, and M. El-Nainay. On stochastic controller placement in software-defined wireless networks. *IEEE Wireless Communications and Networking Conference (WCNC)*, pages 1–6, San Francisco, CA, 2017. DOI: 10.1109/wcnc.2017.7925942. 49, 50

[88] A. Dvir, Y. Haddad, and A. Zilberman. Wireless controller placement problem. *15th IEEE Annual Consumer Communications and Networking Conference (CCNC)*, pages 1–4, Las Vegas, NV, 2018. DOI: 10.1109/ccnc.2018.8319228. 49, 51

[89] M. Johnston and E. Modiano. Controller placement for maximum throughput under delayed CSI. *Modeling and Optimization in Mobile, Ad Hoc, and Wireless Networks (WiOpt), 13th International Symposium on IEEE*, pages 521–528, 2015. DOI: 10.1109/wiopt.2015.7151114. 49, 52, 53, 103

[90] M. Polese, R. Jana, V. Kounev, K. Zhang, S. Deb, and M. Zorzi. Machine learning at the edge: A data-driven architecture with applications to 5G cellular networks. *IEEE Trans. Mobile Comput.*. DOI: 10.1109/tmc.2020.2999852. 49, 53, 54

[91] M. Alharthi, A. M. Taha, and H. S. Hassanein. Dynamic controller placement in software defined drone networks. *IEEE Global Communications Conference (GLOBECOM)*, pages 1–6, Waikoloa, HI, 2019. DOI: 10.1109/GLOBECOM38437.2019.9013799. 49, 55, 56

[92] B. Lantz, B. Heller, and N. McKeown. A network in a laptop: Rapid prototyping for software-defined networks. *Proc. of the 9th ACM SIGCOMM Workshop on Hot Topics in Networks, ACM*, page 19, 2010. DOI: 10.1145/1868447.1868466. 58

[93] Y. E. Oktian, S. Lee, H. Lee, and J. Lam. Distributed SDN controller system: A survey on design choice. *Computer Networks*, 121(5):100–111, 2017. DOI: 10.1016/j.comnet.2017.04.038. 65

[94] A. S. Muqaddas, P. Giaccone, A. Bianco, and G. Maier. Inter-controller traffic to support consistency in ONOS clusters. *IEEE Trans. Netw. Serv. Manage.*, 14(4):1018–1031, 2017. DOI: 10.1109/tnsm.2017.2723477.

[95] S.-C.-C. P. Wang, I. F. Akyildiz, and M. Luo. Towards optimal network planning for software-defined networks. *IEEE Trans. Mobile Comput.*, 2018. DOI: 10.1109/tmc.2018.2815691.

[96] B. Pfaff et al. OpenFlow switch specification v1.3.0. *Tech. Rep.*, 2012. 61

[97] M. Aslan and A. Matrawy. On the impact of network state collection on the performance of SDN applications. *IEEE Commun. Lett.*, 20(1):5–8, January 2016. DOI: 10.1109/lcomm.2015.2496955. 61

[98] A. Yassine, H. Rahimi, and S. Shirmohammadi. Software defined network traffic measurement: Current trends and challenges. *IEEE Instrument. Measure. Mag.*, 18(2):42–50, April 2015. DOI: 10.1109/mim.2015.7066685. 62

[99] P. Tsai, C. Tsai, C. Hsu, and C. Yang. Network monitoring in software-defined networking: A review. *IEEE Syst. J.*, 12(4):3958–3969, December 2018. DOI: 10.1109/JSYST.2018.2798060. 62

[100] N. L. M. van Adrichem, C. Doerr, and F. A. Kuipers. OpenNetMon: Network monitoring in OpenFlow software-defined networks. *IEEE Network Operations and Management Symposium (NOMS)*, pages 1–8, 2014. DOI: 10.1109/noms.2014.6838228. 62

[101] H. Xu, Z. Yu, C. Qian, X. Li, Z. Liu, and L. Huang. Minimizing flow statistics collection cost using wildcard-based requests in SDNs. *IEEE/ACM Trans. Network.*, 25(6):3587–3601, December 2017. DOI: 10.1109/tnet.2017.2748588. 62, 63

[102] A. Tootoonchian, M. Ghobadi, and Y. Ganjali. OpenTM: Traffic matrix estimator for OpenFlow networks. *Proc. 11th Int. Conf. Passive Active Meas.*, pages 201–210, 2010. DOI: 10.1007/978-3-642-12334-4_21. 63

[103] T. Y. Cheng and X. Jia. Compressive traffic monitoring in hybrid SDN. *IEEE Journal on Selected Areas in Communications*, 36(12):2731–2743, December 2018. DOI: 10.1109/jsac.2018.2871311. 62, 64

[104] D. Levin, A. Wundsam, B. Heller, N. Handigol, and A. Feldmann. Logically centralized? State distribution trade-offs in software defined networks. *ACM HotSDN*, 2012. DOI: 10.1145/2342441.2342443. 65, 66

[105] A. Panda, W. Zheng, X. Hu, A. Krishnamurthy, and S. Shenker. SCL: Simplifying distributed SDN control planes. *USENIX NSDI*, 2017. 65

[106] Z. Guo, M. Su, Y. Xu, Z. Duan, L. Wang, S. Hui, and H. J. Chao. Improving the performance of load balancing in software-defined networks through load variance-based synchronization. *Comput. Netw.*, 68:95–109, 2014. DOI: 10.1016/j.comnet.2013.12.004. 66

[107] A. Singla, S. Tschiatschek, and A. Krause. Noisy submodular maximization via adaptive sampling with applications to crowdsourced image collection summarization. *AAAI*, 2016. 69

[108] K. Poularakis, Q. Qin, L. Ma, S. Kompella, K. K. Leung, and L. Tassiulas. Learning the optimal synchronization rates in distributed SDN control architectures. *IEEE International Conference on Computer Communications (Infocom)*, 2019. DOI: 10.1109/infocom.2019.8737388. 65, 66, 68, 69, 70

[109] Z. Zhang, L. Ma, K. Poularakis, K. K. Leung, and L. Wu. DQ scheduler: Deep reinforcement learning based controller synchronization in distributed SDN. *IEEE International Conference on Communications (ICC)*, 2019. DOI: 10.1109/icc.2019.8761183. 65

[110] Z. Zhang, L. Ma, K. Poularakis, K. Leung, J. Tucker, and A. Swami. MACS: Deep reinforcement learning based SDN controller synchronization policy design. *IEEE International Conference on Network Protocols (ICNP)*, 2019. DOI: 10.1109/icnp.2019.8888034. 65

[111] L. Zhao, J. Hua, Y. Liu, W. Qu, S. Zhang, and S. Zhong. Distributed traffic engineering for multi-domain software defined networks. *ICDCS*, 2019. DOI: 10.1109/icdcs.2019.00056. 71, 72

[112] A. Martey. *IS-IS Network Design Solutions*. Cisco Press, Indianapolis, IN, 2002.

[113] M. K. Mukerjee, D. Han, S. Seshan, and P. Steenkiste. Understanding tradeoffs in incremental deployment of new network architectures. *Proc. of ACM CoNEXT*, 2013. DOI: 10.1145/2535372.2535396. 73

[114] S. Vissicchio, L. Vanberer, and O. Bonaventure. Opportunities and research challenges of hybrid software defined networks. *ACM CCR*, 44(2), 2014. DOI: 10.1145/2602204.2602216. 73

[115] Reuters Technology News, AT&T to buy Cisco Core Routers for Network Upgrade. http://www.reuters.com/article/us-cisco-att-idUSN1039414520071210, July 2016.

[116] Z. Cao, M. Kodialam, and T. V. Lakshman. Traffic steering in software defined networks: Planning and online routing. *ACM DCC*, 2014. DOI: 10.1145/2740070.2627574. 74

[117] S. Agarwal, M. Kodialam, and T. V. Lakshman. Traffic engineering in software defined networks. *IEEE Infocom*, 2013. DOI: 10.1109/infcom.2013.6567024. 74

[118] Light Reading Portal, NEC Slashes OpenFlow SDN Controller Pricing. http://www.lightreading.com/carrier-sdn/sdn-technology/nec-slashes-openflow-sdn-controller-pricing/d/d-id/711391, 2014.

[119] K. Poularakis, G. Iosifidis, G. Smaragdakis, and L. Tassiulas. One step at a time: Optimizing SDN upgrades in ISP networks. *IEEE Infocom*, 2017. DOI: 10.1109/infocom.2017.8057136. 76, 78, 79, 82, 83

[120] K. Poularakis, G. Iosifidis, G. Smaragdakis, and L. Tassiulas. Optimizing gradual SDN upgrades in ISP networks. *IEEE/ACM Trans. Network.*, 27(1):288–301, February 2019. DOI: 10.1109/tnet.2018.2890248. 82, 83

[121] E. H.-K. Wu, B. Kar, and Y.-D. Lin. The budgeted maximum coverage problem in partially deployed software defined networks. *IEEE Trans. Netw. Serv. Manage.*, 13(3):394–406, 2016. DOI: 10.1109/tnsm.2016.2598549. 82, 83

[122] X. Jia, Y. Jiang, and Z. Guo. Incremental switch deployment for hybrid software-defined networks. *IEEE LCN*, 2016. DOI: 10.1109/lcn.2016.95. 82, 83

[123] H. Xu, X.-Y. Li, L. Huang, H. Deng, H. Huang, and H. Wang. Incremental deployment and throughput maximization routing for a hybrid SDN. *IEEE/ACM Trans. Networking*, 2017. DOI: 10.1109/tnet.2017.2657643. 82, 83

[124] L. Wang, Q. Li, Y. Jiang, and J. Wu. Towards mitigating link flooding attack via incremental SDN deployment. *IEEE ISCC*, 2016. DOI: 10.1109/ISCC.2016.7543772. 82, 83

[125] M. Caria, A. Jukan, and M. Hoffmann. Divide and conquer: Partitioning OSPF networks with SDN. *IEEE IM*, 2015. DOI: 10.1109/inm.2015.7140324. 82, 83

[126] M. Caria, A. Jukan, and M. Hoffmann. A performance study of network migration to SDN-enabled traffic engineering. *IEEE Globecom*, 2013. DOI: 10.1109/glocom.2013.6831268. 82, 83

[127] T. Das, M. Caria, A. Jukan, and M. Hoffmann. A techno-economic analysis of network migration to software-defined networking. *ArXiv. 1310.0216v1*, 2013. 82, 83

[128] D. Levin, M. Canini, S. Schmid, F. Schaffert, and A. Feldmann. Panopticon: Reaping the benefits of incremental SDN deployment in enterprise networks. *Proc. of USENIX ATC*, 2014. 82, 83

[129] D. K. Hong, Y. Ma, S. Banerjee, and Z. M. Mao. Incremental deployment of SDN in hybrid enterprise and ISP networks. *ACM SOSR*, 2016. DOI: 10.1145/2890955.2890959. 82, 83

[130] H. Xu, J. Fan, J. Wu, C. Qiao, and L. Huang. Joint deployment and routing in hybrid SDNs. *IEEE/ACM IWQoS*, 2017. DOI: 10.1109/IWQoS.2017.7969133. 82, 83

[131] S. Vissicchio, O. Tilmans, L. Vanbever, and J. Rexford. Central control over distributed routing. *Prof. of ACM SIGCOMM*, 2015. DOI: 10.1145/2785956.2787497. 85

[132] A. Cianfrani, M. Listanti, and M. Polverini. Incremental deployment of segment routing into an ISP network: A traffic engineering perspective. *IEEE/ACM Trans. Networking*, 2017. DOI: 10.1109/tnet.2017.2731419. 86

[133] M. Pioro and D. Mehdi. *Routing, Flow, and Capacity Design in Communication and Computer Networks*. Morgan Kaufmann Publishers, 2004. DOI: 10.1016/B978-0-12-557189-0.X5000-8. 87

[134] S. Agarwal, M. Kodialam, and T. V. Lakshman. Traffic engineering in software defined networks. *Infocom*, 2013. DOI: 10.1109/infcom.2013.6567024. 88, 90

[135] N. Buchbinder and J. Naor. Improved bounds for online routing and packing via a primal-dual approach. *47th Annual IEEE Symposium on Foundations of Computer Science (FOCS'06)*, pages 293–304, 2006. DOI: 10.1109/focs.2006.39. 91

[136] L. Luo, H. Yu, Z. Ye, and X. Du. Online deadline-aware bulk transfer over inter-datacenter WANs. *INFOCOM—IEEE Conference on Computer Communications*, pages 630–638, 2018. DOI: 10.1109/infocom.2018.8485828. 91

[137] O. Alhussein and W. Zhuang. Robust online composition, routing and NF placement for NFV-enabled services. *IEEE J. Select. Areas Commun.*, 38(6):1089–1101, June 2020. DOI: 10.1109/jsac.2020.2986612. 91

[138] M. Huang, W. Liang, Z. Xu, W. Xu, S. Guo, and Y. Xu. Dynamic routing for network throughput maximization in software-defined networks. *INFOCOM—The 35th Annual IEEE International Conference on Computer Communications*, pages 1–9, 2016. DOI: 10.1109/infocom.2016.7524613. 91

[139] N. Liakopoulos, A. Destounis, G. Paschos, T. Spyropoulos, and P. Mertikopoulos. Cautious regret minimization: Online optimization with long-term budget constraints. *ICML*, 2019. 92

[140] M. Rifai et al. Too many SDN rules? Compress them with MINNIE. *IEEE Global Communications Conference (GLOBECOM)*, pages 1–7, 2015. DOI: 10.1109/GLOCOM.2015.7417661. 94

[141] R. Cohen, L. Lewin-Eytan, J. S. Naor, and D. Raz. On the effect of forwarding table size on SDN network utilization. *INFOCOM—IEEE Conference on Computer Communications*, pages 1734–1742, 2014. DOI: 10.1109/infocom.2014.6848111. 93

[142] X. Nguyen, D. Saucez, C. Barakat, and T. Turletti. OFFICER: A general optimization framework for OpenFlow rule allocation and endpoint policy enforcement. *IEEE Conference on Computer Communications (INFOCOM)*, pages 478–486, 2015. DOI: 10.1109/infocom.2015.7218414. 93

[143] B. Heller, S. Seetharaman, P. Mahadevan, Y. Yiakoumis, P. Sharma, S. Banerjee, and N. McKeown. ElasticTree: Saving energy in data center networks. *NSDI*, 2010. 94

[144] P. T. Congdon, P. Mohapatra, M. Farrens, and V. Akella. Simultaneously reducing latency and power consumption in OpenFlow switches. *IEEE/ACM Trans. Netw.*, 22(3):1007–1020, June 2014. DOI: 10.1109/tnet.2013.2270436. 94

[145] S. Jouet, C. Perkins, and D. Pezaros. OTCP: SDN-managed congestion control for data center networks. *NOMS—IEEE/IFIP Network Operations and Management Symposium*, pages 171–179, Istanbul, 2016. DOI: 10.1109/noms.2016.7502810. 98

[146] J. Perry, A. Ousterhout, H. Balakrishnan, D. Shah, and H. Fugal. Fastpass: A centralized "zero-queue" data center network. *Proc. of the ACM Conference on SIGCOMM (SIGCOMM'14). Association for Computing Machinery*, 307–318, New York, 2014. DOI: 10.1145/2619239.2626309. 99

[147] M. Al-Fares, S. Radhakrishnan, B. Raghavan, N. Huang, and A. Vahdat. Hedera: Dynamic flow scheduling for data center networks. *NSDI*, 2010. 99

[148] B. C. Vattikonda, G. Porter, A. Vahdat, and A. C. Snoeren. Practical TDMA for datacenter ethernet. *EuroSys*, 2012. DOI: 10.1145/2168836.2168859. 99

[149] J. Perry, H. Balakrishnan, and D. Shah. Flowtune: Flowlet control for datacenter networks. *14th USENIX Symposium on Networked Systems Design and Implementation (NSDI'17). USENIX Association*, Boston, MA, 2017. 99

[150] L. Liu, D. Li, and J. Wu. TAPS: Software defined task-level deadline-aware preemptive flow scheduling in data centers. *44th International Conference on Parallel Processing*, pages 659–668, Beijing, 2015. DOI: 10.1109/icpp.2015.75. 99

[151] R. Bhatia, F. Hao, M. Kodialam, and T. V. Lakshman. Optimized network traffic engineering using segment routing. *IEEE Conference on Computer Communications (INFOCOM)*, pages 657–665, Kowloon, 2015. DOI: 10.1109/infocom.2015.7218434. 95, 96, 97

[152] P. L. Ventre, S. Salsano, M. Polverini, A. Cianfrani, A. Abdelsalam, C. Filsfils, P. Camarillo, and F. Clad. Segment routing: A comprehensive survey of research activities, standardization efforts and implementation results. *ArXiv:1904.03471*, 2020. DOI: 10.1109/comst.2020.3036826. 97

[153] L. Tassiulas and A. Ephremides. Stability properties of constrained queueing systems and scheduling policies for maximum throughput in multihop radio networks. *29th IEEE Conference on Decision and Control*, 4:2130–2132, 1990. DOI: 10.1109/CDC.1990.204000. 102, 103

[154] Z. Jiao, B. Zhang, C. Li, and H. T. Mouftah. Backpressure-based routing and scheduling protocols for wireless multihop networks: A survey. *IEEE Wireless Commun.*, 23(1):102–110, February 2016. DOI: 10.1109/mwc.2016.7422412. 103

[155] A. Sinha and E. Modiano. Optimal control for generalized network-flow problems. *IEEE/ACM Trans. Network.*, 26(1):506–519, February 2018. DOI: 10.1109/TNET.2017.2783846. 104

[156] Q. Liang and E. Modiano. Optimal network control in partially-controllable networks. *INFOCOM—IEEE Conference on Computer Communications*, pages 397–405, 2019. DOI: 10.1109/infocom.2019.8737528. 104

[157] N. Alliance. NGMN 5G P1 requirements and architecture work stream end-to-end architecture description of network slicing concept. https://www.ngmn.org/uploads/media/161010 010 NGMN Network Slicing framework v1.0.8.pdf, 2016.

[158] M. Liu, A. W. Richa, M. Rost, and S. Schmid. A constant approximation for maximum throughput multicommodity routing and its application to delay-tolerant network scheduling. *INFOCOM*, pages 46–54, 2019. DOI: 10.1109/infocom.2019.8737402. 110

[159] R. Cohen, L. Lewin-Eytan, J. S. Naor, and D. Raz. Near optimal placement of virtual network functions. *IEEE Infocom*, 2015. DOI: 10.1109/infocom.2015.7218511. 110

[160] I. Benkacem, T. Taleb, M. Bagaa, and H. Flinck. Optimal VNFs placement in CDN slicing over multi-cloud environment. *IEEE J. Select. Areas Commun.*, 36(3):616–627, 2018. DOI: 10.1109/jsac.2018.2815441. 110

[161] A. Laghrissi, T. Taleb, and M. Bagaa. Conformal mapping for optimal network slice planning based on canonical domains. *IEEE J. Select. Areas Commun.*, 36(3):519–528, 2018. DOI: 10.1109/jsac.2018.2815436. 110

[162] M. Rost and S. Schmid. Virtual network embedding approximations: Leveraging randomized rounding. *IFIP Networking*, 2018. 110, 112

[163] M. A. T. Nejad, S. Parsaeefard, M. A. Maddah-Ali, T. Mahmoodi, and B. H. Khalaj. vSPACE: VNF simultaneous placement, admission control and embedding. *IEEE J. Select. Areas Commun.*, 36(3):542–557, 2018. DOI: 10.1109/jsac.2018.2815318. 110

[164] J. Pei, P. Hong, K. Xue, and D. Li. Efficiently embedding service function chains with dynamic virtual network function placement in geo-distributed cloud system. *IEEE Trans. Parallel Distrib. Syst.*, 2019. DOI: 10.1109/tpds.2018.2880992. 110

[165] B. Addis, D. Belabed, M. Bouet, and S. Secci. Virtual network functions placement and routing optimization. *Cloudnet*, 2015. DOI: 10.1109/cloudnet.2015.7335301. 110

[166] M. Barcelo, A. Correa, J. Llorca, A. M. Tulino, J. L. Vicario, and A. Morell. IoT-cloud service optimization in next generation smart environments. *IEEE J. Select. Areas Commun.*, 2016. DOI: 10.1109/jsac.2016.2621398. 110

[167] J. Liu, W. Lu, F. Zhou, P. Lu, and Z. Zhu. On dynamic service function chain deployment and readjustment. *IEEE Trans. Netw. Serv. Manage.*, 14(3):543–553, 2017. DOI: 10.1109/tnsm.2017.2711610. 110

[168] C. Pham, N. H. Tran, S. Ren, W. Saad, and C. S. Hong. Traffic-aware and energy efficient VNF placement for service chaining: Joint sampling and matching approach. *IEEE Trans. Serv. Comput.*, 2017. DOI: 10.1109/tsc.2017.2671867. 110

[169] S. Agarwal, F. Malandrino, C. F. Chiasserini, and S. De. VNF placement and resource allocation for the support of vertical services in 5G networks. *IEEE/ACM Trans. Network.*, 2019. DOI: 10.1109/tnet.2018.2890631. 110

[170] I. Baev, R. Rajaraman, and C. Swamy. Approximation algorithms for data placement problems. *SIAM J. Comp.*, 38, 2008. DOI: 10.1137/080715421. 110

[171] S. Borst, V. Gupta, and A. Walid. Distributed caching algorithms for content distribution networks. *IEEE Infocom*, 2010. DOI: 10.1109/infcom.2010.5461964. 110

[172] K. Shanmugam, N. Golrezaei, A. Dimakis, A. Molisch, and G. Caire. FemtoCaching: Wireless content delivery through distributed caching helpers. *IEEE Trans. Inform. Theor.*, 59(12), 2013. DOI: 10.1109/tit.2013.2281606. 110

[173] T. He, H. Khamfroush, S. Wang, T. L. Porta, and S. Stein. It's hard to share: Joint service placement and request scheduling in edge clouds with sharable and non-sharable resources. *IEEE ICDCS*, 2018. DOI: 10.1109/icdcs.2018.00044. 110

[174] J. Xu, L. Chen, and P. Zhou. Joint service caching and task offloading for mobile edge computing in dense networks. *IEEE Infocom*, 2018. DOI: 10.1109/INFO-COM.2018.8485977. 110

[175] K. Poularakis, J. Llorca, A. Tulino, I. Taylor, and L. Tassiulas. Joint service placement and request routing in multi-cell mobile edge computing networks. *IEEE Infocom*, 2019. DOI: 10.1109/infocom.2019.8737385. 110

[176] K. Poularakis, J. Llorca, A. Tulino, and L. Tassiulas. Approximation algorithms for data-intensive service chain embedding. *ACM Mobihoc*, 2020. DOI: 10.1145/3397166.3409149. 111, 114

[177] T. Hurley, J. E. Perdomo, and A. Perez-Pons. HMM-based intrusion detection system for software defined networking. *Proc. IEEE ICMLA*, pages 617–621, Anaheim, CA, December 2016. DOI: 10.1109/icmla.2016.0108. 117

[178] T. A. Tang, L. Mhamdi, D. McLernon, S. A. R. Zaidi, and M. Ghogho. Deep learning approach for network intrusion detection in software defined networking. *Proc. IEEE WINCOM*, pages 258–263, Fes, Morocco, October 2016. DOI: 10.1109/wincom.2016.7777224. 117

[179] T. Tang, S. A. R. Zaidi, D. McLernon, L. Mhamdi, and M. Ghogho. Deep recurrent neural network for intrusion detection in SDN-based networks. *Proc. IEEE NetSoft*, pages 1–5, Montreal, QC, Canada, 2018. DOI: 10.1109/netsoft.2018.8460090. 117

[180] N. Napiah, M. Y. I. B. Idris, R. Ramli, and I. Ahmedy. Compression header analyzer intrusion detection system (cha-ids) for 6lowpan communication protocol. *IEEE Access*, 6:16 623–16 638, 2018. DOI: 10.1109/access.2018.2798626. 117

[181] W. Wang, M. Zhu, J. Wang, X. Zeng, and Z. Yang. End-to-end encrypted traffic classification with one-dimensional convolution neural networks. *IEEE International Conference on Intelligence and Security Informatics (ISI)*, pages 43–48, 2017. DOI: 10.1109/isi.2017.8004872. 117

[182] M. Lotfollahi, M. J. Siavoshani, R. S. H. Zade, and M. Saberian. Deeppacket: A novel approach for encrypted traffic classification using deep learning. *Soft Computing*, pages 1–14, 2017. DOI: 10.1007/s00500-019-04030-2. 117

[183] Z. Wang. The applications of deep learning on traffic identification. *BlackHat*, 24, 2015. 117

[184] Q. Qin, K. Poularakis, and L. Tassiulas. A learning approach with programmable data plane towards IoT security. *IEEE International Conference on Distributed Computing Systems (ICDCS)*, 2020. DOI: 10.1109/icdcs47774.2020.00064. 117, 118

[185] A. Capone, C. Cascone, A. Q. T. Nguyen, and B. Sansò. Detour planning for fast and reliable failure recovery in SDN with OpenState. *11th International Conference on the Design of Reliable Communication Networks (DRCN)*, pages 25–32, 2015. DOI: 10.1109/drcn.2015.7148981. 121

[186] X. Zhang, L. Cui, K. Wei, F. P. Tso, Y. Ji, and W. Jia. A survey on stateful data plane in software defined networks. *Comput. Netw.*, page 107597, 2020. DOI: 10.1016/j.comnet.2020.107597. 121

[187] C. Kim, A. Sivaraman, N. Katta, A. Bas, A. Dixit, and L. J. Wobker. In-band network telemetry via programmable data planes. *ACM SIGCOMM Symposium on SDN Research (SOSR)*, 2015. 124

[188] T. Pan et al. INT-path: Towards optimal path planning for in-band network-wide telemetry. *INFOCOM—IEEE Conference on Computer Communications*, pages 487–495, 2019. DOI: 10.1109/infocom.2019.8737529. 124

[189] X. Yu, H. Xu, D. Yao, H. Wang, and L. Huang. CountMax: A lightweight and cooperative sketch measurement for software-defined networks. *IEEE/ACM Trans. Network.*, 26(6):2774–2786, December 2018. DOI: 10.1109/tnet.2018.2877700. 126

[190] Y. Zhai, H. Xu, H. Wang, Z. Meng, and H. Huang. Joint routing and sketch configuration in software-defined networking. *IEEE/ACM Trans. Network.*, 28(5):2092–2105, October 2020. DOI: 10.1109/tnet.2020.3002783. 126, 127

[191] H. Wang and B. Li. Lube: Mitigating bottlenecks in wide area data analytics. *HotCloud*, 2017. 131

[192] A. Valadarsky, M. Schapira, D. Shahaf, and A. Tamar. Learning to route. *Proc. ACM HotNets*, pages 185–191, 2017. DOI: 10.1145/3152434.3152441. 131, 132

[193] H. Mao, M. Alizadeh, I. Menache, and S. Kandula. Resource management with deep reinforcement learning. *HotNets*, 2016. DOI: 10.1145/3005745.3005750. 131

[194] K. Winstein and H. Balakrishnan. TCP ex Machina: Computer-generated congestion control. *SIGCOMM*, 2013. DOI: 10.1145/2534169.2486020. 131

[195] H. Mao, R. Netravali, and M. Alizadeh. Neural adaptive bitrate streaming with pensive. *SIGCOMM*, 2017. DOI: 10.1145/3098822.3098843. 131

[196] J. Xie, F. R. Yu, T. Huang, R. Xie, J. Liu, and Y. Liu. A survey of machine learning techniques applied to software defined networking (SDN): Research issues and challenges. *IEEE Commun. Surv. Tutor.* , 2018. DOI: 10.1109/comst.2018.2866942. 131

[197] R. Muñoz et al. Integrated SDN/NFV management and orchestration architecture for dynamic deployment of virtual SDN control instances for virtual tenant networks. *J. Opt. Commun. Netw.*, 7(11):62–70, 2015. DOI: 10.1364/jocn.7.000b62.

[198] Q. Qin, K. Poularakis, K. K. Leung, and L. Tassiulas. Line-speed and scalable intrusion detection at the network edge via federated learning. *IFIP Networking*, 2020. 133

[199] X. Jin et al. SoftCell: Taking control of cellular core networks. *Proc. ACM CoNEXT*, pages 1–14, 2013. 133

[200] A. Gudipati et al. SoftRAN: Software defined radio access network. *Proc. ACM HotSDN*, pages 25–30, 2013. DOI: 10.1145/2491185.2491207. 133

[201] M. Bansal et al. Openradio: A programmable wireless data plane. *Proc. ACM HotSDN*, pages 109–14, 2012. DOI: 10.1145/2342441.2342464. 133

[202] J. Lee et al. meSDN: Mobile extension of SDN. *Proc. 5th Int. Workshop Mobile Cloud Comput. Services*, pages 7–14, Bretton Woods, NH, 2014. DOI: 10.1145/2609908.2609948. 133

[203] K. Poularakis, Q. Qin, K. M. Marcus, K. S. Chan, K. K. Leung, and L. Tassiulas. Hybrid SDN control in mobile ad hoc networks. *IEEE Workshop on Distributed Analytics Infrastructure and Algorithms for Multi-Organization Federations, in Proc. of IEEE Smart Computing*, 2019. DOI: 10.1109/smartcomp.2019.00038. 133

[204] ARL project. Dais ITA: The distributed analytics and information science international technology alliance, 2016–2021. 65, 133

Authors' Biographies

KONSTANTINOS POULARAKIS

Konstantinos Poularakis is a Research Scientist at Yale University. His research interests lie in the area of network optimization and machine learning with emphasis on emerging architectures such as software-defined networks, mobile edge computing, and caching networks. His research has been recognized by several awards and scholarships from sources including the Greek State Scholarships foundation (2011), the Center for Research and Technology Hellas (2012), the Alexander S. Onassis Public Benefit Foundation (2013), and the Bodossaki Foundation (2016). He also received the Best Paper Award at the IEEE Infocom 2017 and IEEE ICC 2019 conferences. Konstantinos obtained his Diploma, M.S., and Ph.D. degrees in Electrical Engineering from the University of Thessaly, Greece, in 2011, 2013, and 2016 respectively. In 2014, he was a Research Intern with Technicolor Research, Paris.

LEANDROS TASSIULAS

Leandros Tassiulas is the John C. Malone Professor of Electrical Engineering at Yale University. His research interests are in the field of computer and communication networks with emphasis on fundamental mathematical models and algorithms of complex networks, architectures, and protocols of wireless systems, sensor networks, novel internet architectures, and experimental platforms for network research. His most notable contributions include the max-weight scheduling algorithm and the back-pressure network control policy, opportunistic scheduling in wireless, the maximum lifetime approach for wireless network energy management, and the consideration of joint access control and antenna transmission management in multiple antenna wireless systems. Dr. Tassiulas is a Fellow of IEEE (2007) and the ACM (2020). His research has been recognized by several awards including the IEEE Koji

Kobayashi Computer and Communications Award (2016), the ACM SIGMETRICS Achievement Award 2020, the inaugural INFOCOM 2007 Achievement Award "for fundamental contributions to resource allocation in communication networks," several best paper awards including the at INFOCOM 1994, and 2017 and Mobihoc 2016, a National Science Foundation (NSF) Research Initiation Award (1992), an NSF CAREER Award (1995), an Office of Naval Research Young Investigator Award (1997), and a Bodossaki Foundation Award (1999). He holds a Ph.D. in Electrical Engineering from the University of Maryland, College Park (1991) and a Diploma in Electrical Engineering from Aristotle University of Thessaloniki (1987). He has held faculty positions at Polytechnic University, New York, University of Maryland, College Park, and University of Thessaly, and University of Ioannina, Greece.

T.V. LAKSHMAN

T.V. Lakshman is the Head of the Networks Research Group, Nokia Bell Laboratories. His research contributions to networking span a spectrum of topics, including software-defined networking, traffic management, switch architectures, network optimization, TCP performance, and machine-learning applications to networks. He is a recipient of several IEEE and ACM awards, including the IEEE Leonard Abraham Prize, the IEEE Communication Society William R. Bennett Prize, the IEEE INFOCOM Achievement Award, the IEEE Fred W. Ellersick Prize Paper Award, the ACM SIGMETRICS Best Paper Award, and the IEEE Infocom Best Paper Award. He also received the 2010 Thomas Alva Edison Patent Award from the R&D Council of New Jersey. He has been an Editor of the *IEEE/ACM Transactions on Networking* and the *IEEE Transactions on Mobile Computing*. He is a Fellow of IEEE, ACM, and Bell Labs. He received the Master's degree (Physics) from the Indian Institute of Science, Bengaluru, India, and the M.S. and Ph.D. degrees in Computer Science from the University of Maryland, College Park.